AF774989

Ancient Christian Ecopoetics

DIVINATIONS: REREADING LATE ANCIENT RELIGION

Series Editors: Daniel Boyarin, Virginia Burrus, Derek Krueger

A complete list of books in the series
is available from the publisher.

ANCIENT CHRISTIAN ECOPOETICS

COSMOLOGIES, SAINTS, THINGS

VIRGINIA BURRUS

PENN

UNIVERSITY OF PENNSYLVANIA PRESS

PHILADELPHIA

Copyright © 2019 University of Pennsylvania Press

All rights reserved. Except for brief quotations used for purposes of review
or scholarly citation, none of this book may be reproduced in any form
by any means without written permission from the publisher.

Published by
University of Pennsylvania Press
Philadelphia, Pennsylvania 19104-4112
www.upenn.edu/pennpress

Printed in the United States of America
on acid-free paper

1 3 5 7 9 10 8 6 4 2

Library of Congress Cataloging-in-Publication Data

Names: Burrus, Virginia, author.
Title: Ancient Christian ecopoetics: cosmologies, saints, things / Virginia Burrus.
Other titles: Divinations.
Description: 1st edition. | Philadelphia: University of Pennsylvania Press, [2019] | Series: Divinations: rereading late ancient religion | Includes bibliographical references and index.
Identifiers: LCCN 2018015361 | ISBN 9780812250794 (hardcover: alk. paper)
Subjects: LCSH: Ecology—Religious aspects—Christianity. | Church history—Primitive and early church, ca. 30-600. | Cosmology, Ancient. | Christian hagiography—History—To 1500. | Material culture—Religious aspects.
Classification: LCC BT695.5 .B855 2018 | DDC 230/.13—dc23
LC record available at https://lccn.loc.gov/2018015361

For Catherine and Sharon,
beloved sister-friends,
without whom this book would have been unthinkable

Contents

INTRODUCTION

> We are only just beginning to think the ecological thought. Perhaps there is no end to its thinking.
>
> —Timothy Morton

What does it mean *to think the ecological thought*? For Timothy Morton and a host of other ecologically minded theorists, it entails attentiveness to the interconnection and the liveliness and agency of all beings. If some of us are just beginning to think this thought, this is because we are just beginning to comprehend the consequences of the failure to do so. Having believed ourselves masters of nature, we now learn the painful lessons of dependence and vulnerability. Ecological thought comes belatedly then. It arrives only when it is too late to undo the planetary damage. Yet thinking it matters more than ever. It is also more than ever within our grasp. Indeed, crisis births revelation, and apparent end times may turn out to offer "the first glimmerings of new times."[1]

Ecological thought propels us forward toward a future as yet barely imaginable. Paradoxically, it may also draw us back to a past only dimly recalled. What draws us, however, is not the longing for a simpler, purer time and place. Even if such existed, it would be of little help to us now. Rather, the pursuit of a usable past here evokes a context as complex and in its own way as compromised as our own—namely, the late ancient Mediterranean. The cultural artifacts of Christianity in particular claim our attention. What might they have to say to us "anthropocenes"—we who have outlived our innocence, no longer able to imagine our planet impervious to destructive human habit?[2]

To some, this question may seem odd or forced. Why should we expect ancient Christian theologies, stories, images, or practices to speak to ecological crisis, or indeed to any of the other urgencies of our own moment? Why should we even want them to? As Christopher Schliephake notes in his introduction to a volume titled *Ecocriticism, Ecology, and the Cultures of Antiquity*, there is danger in hitching our historical studies so tightly to issues of acknowledged contemporary relevance that we cease to be challenged and provoked by the alien character of the past. "It is one of the (hopefully) enduring achievements of a humanistic education that the study of worlds far removed from our own has value in itself."[3] Yet of course we always walk a fine line: an escapist history is no more alluring than an instrumentalized history. Moreover, there are many readers for whom my question will seem neither odd nor forced. Christianity has often been seen as a culprit—if not *the* culprit—in the history of human-induced environmental degradation. What Roman imperialism began, with its strategies of territorial conquest and exploitation, Christian monotheism finished, in the view of environmental historian Donald Hughes: monotheism "taught that God was separate from creation, and denied any inherent sacredness in nature"; moreover, "the first book of the Bible said that God had given man 'dominion' over the Earth."[4] Less often, perhaps, Christianity has been looked to as a possible savior: "The notion that the Logos can be seen in every created thing—that the world is in some sense a living museum of divine intent—is scandalously powerful," enthuses environmental activist and journalist Bill McKibben in the foreword to a volume of essays that seeks ecological resources within the Eastern Orthodox tradition.[5]

Thus my question about the possible relevance of ancient Christianity to current ecological thought has a larger context and a longer history. One starting point is medieval historian Lynn White's 1967 essay "The Historical Roots of Our Ecological Crisis." Invoking the histories of technology, science, politics, culture, and religion, White's claims in this five-page manifesto are bold and multifaceted. Three in particular have incited discussion and debate in the half-century since the article was published, as Todd LeVasseur and Anna Peterson observe in a volume dedicated to reflection on that fifty-year marker.[6] First, White proposes that how humans behave in relation to the world around us "is deeply conditioned by beliefs about our nature and destiny—that is, by religion."[7] Second, he singles out Christianity—especially "Western" Christianity—as particularly implicated in the current ecological crisis owing to its strident anthropocentrism and its consequent tendency to devalue the nonhuman world, attitudes conveyed by a creation narrative that insinuates that humans are uniquely formed in the image of a god who grants them dominion over a cosmos existing only to serve them. Third, White suggests that "since the roots of our trouble are so largely religious, the remedy must also be essentially religious, whether we call it that or not." Pursuing such a remedy, he turns to Saint Francis for "an alternative Christian view of nature and man's relation to it," noting that Francis "tried to substitute the idea of the equality of all creatures, including man, for the idea of man's limitless rule of creation."[8]

All three of these claims have been vigorously contested, yet none has been simply discarded. First, how people think about the world around them—whether we call it worldview, cosmology, theology, or ideology—is not the *only* factor affecting their behavior toward that more-than-human universe, most would agree. It may not even be the most important factor much of the time. As Willis Jenkins points out, "The ethicist might entertain the converse to White's thesis: why not hold that cosmology ('what they think about themselves in relation to the things around them') is produced by social practice ('what people do about their ecology')? If that were the case, it would shift the ethical task away from transforming cosmology and toward transforming social practices."[9] Affect theory in particular has illumined the extent to which humans, like other animals, are largely moved by embodied histories, forces, and relations that emerge "before language, before cosmology, even before 'thought,'" as Donovan Schaefer argues. Yet thought is not always merely "a way of converting a situation into an explanation," as Schaefer suggests.[10] Thought may also shift, mute, or intensify affects; it may reshape practices. In other words, how humans imagine, explain, narrate, or represent their wider worlds matters in and for those worlds, whether confirming or challenging what are often deep-seated habits of behavior and relation; thus it matters too that we reflect critically on such practices of imagination, explanation, narration, or representation. "If we want a good reality—say, for instance, nonviolent coexistence between all beings—we might need to figure out what kinds of attitude are conducive to such a reality," as Morton puts it.[11]

Second, White's view of Christianity strikes many as troublingly monolithic and essentializing. At the same time, his critique of a "Western" Christianity crystallized in the European Middle Ages may seem too narrowly targeted, and his consequent idealizing of Eastern Christianity and Asian religions has opened him to the charge of a romanticizing orientalism.[12] Nonetheless, his particular criticisms of Christianity, however much painted in broad strokes, have largely stuck. Christianity's anthropocentrism, its eschatologically refracted dualism, and its theology of dominion have often hindered rather than enhanced ecologically sensitive thought and action.

Third, few would agree that there is a single remedy for our "trouble," much less that the remedies will always be "religious" in any sense, much less still that they are or should be distinctly Christian, as White may seem to imply. At the same time, the field of Christian ecotheology has burgeoned in the years since White's essay appeared, as scholars and activists have responded with both creativity and passion to the call to render theology more ecologically resonant, in part by drawing on marginal or overlooked strands of their own tradition. Feminist theologians, such as Rosemary Radford Ruether, Sallie McFague, and Ivone Gebara (to name only some of the earliest and most influential voices), have made particularly incisive contributions, at once critical and constructive. In recent years, ecotheology may seem to have become virtually mainstream: Bartholemew, ecumenical patriarch since 1991, has consistently championed ecological concerns, and Pope Francis has now joined his voice to the ecologically minded chorus with the papal encyclical "Laudato Si" of 2013. Among Christian ecotheological works especially germane to this present study is Douglas Christie's *The Blue Sapphire of the Mind*, which turns to late ancient monastic traditions to articulate a "contemplative ecology."[13] Meanwhile, the conversation has also become much more religiously pluralistic, as perhaps evidenced most concretely by the Forum on Religion and Ecology at Yale University, directed by John Grim and Mary Evelyn Tucker.[14] Even an ecocritic working well outside the field of religion is ready to detect hints of hope in "the new eco-religions."[15]

Thus this book situates itself in the trajectory of a now wide-ranging engagement of the fields of religion and ecology that can be seen, at least with hindsight, to have begun with Lynn White's essay. As Whitney Bauman argues, the story could also be told differently: "The White narrative can stand side by side with other narratives that connect that planet in a whole field of intersectional discourses that are working toward a different, more ecologically sound and just planetary future."[16] If I nonetheless begin with "the White narrative," it is so as to *begin again*—to renew and complicate this particular Christian-centered starting point. Ecological critiques of Christian tradition are enormously important but are also by now well established; these will not be my primary focus. Nor am I interested either in defending the "mainstream" tradition (as some of White's critics have done)[17] or in turning to "marginal" ones for leverage against it (as White himself advocated).[18] I want to resist the temptation to look for either a culprit or a savior in Christianity, in other words.

What then am I looking for when I turn back to early Christianity in an effort "to think the ecological thought"? My aims are constructive, but they are neither apologetic nor prescriptive. My discourse is theological, but most of my conversation partners are not theologians; many are not even engaged in the study of religion per se. Bringing current questions, concerns, and theories into dialogue with late ancient Mediterranean ones, I hope to facilitate a provocative encounter with difference, and one that takes poetic form in the broadest sense. As Scott Knickerbocker notes, "A poet crafts language that, if successful, inspires, startles, or coaxes us into knowing the world with revivified senses." "Sensuous poiesis," as Knickerbocker names it, effects an "immediate, embodied experience of nonhuman nature," pointing in the direction of an "ecopoetics" that is undergirded by "the assumption that the same imaginative and intellectual muscles we exercise in our deep consideration of poetry are needed in meaningfully relating to nature and responding to environmental dilemmas—and vice versa."[19] What is true for poetry in the narrow sense is also true for poetry writ large, that is, for all of the ways that *human* creativity both participates in and reflects upon what ancient Christians would have thought of as *divine* creativity or creation. Theology is a kind of poetry, or "making," as advocates of "theopoetics" have long suggested; so too is hagiography or pictorial art. These forms of poiesis both do and do not stand apart from the sacred poiesis that we might, for better or worse, also call nature. Here I am not appealing to a notion of the divine inspiration of theology, literature, or art, nor do I mean to align such human creativity with the transcendently "natural." Rather, I want to suggest, with David L. Miller, that theology, literature, and art may "refer to strategies of human signification in the absence of fixed and ultimate meanings accessible to knowledge or faith."[20] Such creaturely poiesis is productive and performative rather than referential, representational, or propositional; in this, it joins the ongoing processes of becoming that constitute the very universe, while also drawing our attention to those processes through its distinctive reflexivity. Ecopoetics both reflects on and takes part in "the emergence of new forms of life," as Jonathan Skinner puts it.[21]

Ancient Christianity continues to be deemed normative or authoritative for Christians worldwide, and thus many tend to discover their own image in its thought, literature, and art. It may therefore seem an unlikely place to discover the emergence of the new. Yet the world of the so-called Church Fathers is surely far stranger than we tend to imagine. I want to explore the strangeness of ancient Christianity, allowing myself to be inspired, startled, or coaxed by its alien character. Here at the beginning, let me name a few basic facets of that strangeness, as I perceive it.

First, the boundaries between Christianity and what we ourselves might think of as its religious "others" were not always as evident to denizens of the ancient Mediterranean as they are to us. This ancient ecopoetics is all the more Christian because it is also Jewish and Platonic and polytheistic.

Second, ancient Christianity did not yet know itself as Protestant or Catholic or Orthodox, or even as Western or Eastern. Its differentiations—its competing orthodoxies and heresies—were less trenchantly institutionalized and acculturated, more fluid and dynamic (if often no less violent) in their ongoing negotiations than such labels might imply. This ancient ecopoetics leaves the definition of what is properly Christian open.

Third, ancient Mediterranean Christians did not oppose science to belief, reason to faith. Their cosmologies were creative exercises of intellect and imagination, at once scriptural and philosophical, pushing the boundaries of what thought could think. This ancient ecopoetics resorts to dogma only in the root sense of exploratory opinion, to doctrine only in the root sense of teaching.

Fourth, ancient Mediterranean Christians were drawn to excess and transgression, in search of transformation. Pushing against the limits of their very humanity, they eschewed conventional social, sexual, and gender roles and relations, or fantasized about others who did so. This ancient ecopoetics is both queer and posthuman—which is also to say prehumanistic.

Fifth, ancient Mediterranean Christians experienced humans as coexisting with a wide range of other lively, relational beings, whether animal, vegetable, or mineral, angelic or demonic, divine or creaturely, large or small. This ancient ecopoetics is both materialist and animistic.

Of course, what is strange to me may seem familiar to you, just as you may find my familiar things strange. I describe the alien past as I encounter it. I hope to hold onto a sense of its strangeness while at the same time rendering it more intimately familiar—and, yes, more relevant as well. Perhaps it is inevitable that I too will in the end simply have discovered in ancient Christianity the image of my own fears and hopes. But will it be exactly the same image with which I began?

* * *

Cosmologies. Saints. Things. This book unfolds in three parts, each of which is composed of a series of fragments of varying lengths. Each part is introduced by a theoretical "Prelude," while a penultimate section, labeled "In/

Conclusion," draws together the themes and arguments of the whole. Other fragments—"Interludes" and "Postludes"—are both briefer and more experimental in voice and style. The reader can engage each of the parts and to an extent each of the fragments within the parts either as separate units or as components of a larger, if still fragmentary, collectivity. She can also read the parts in different orders. Rather than progressing in a strict sequence, the argument accumulates; it gains resonance, or so I hope, as the reader discovers or creates connections among the fragments. Here form follows content, partly disrupting both linearity and closure: ecopoiesis, as Skinner describes it, "remain[s] open to the incompleteness of its own organization."[22]

Part I considers cosmology under the sign of the Platonic *khora*—the mysterious "third kind" that eludes both intellect and senses, variously dubbed "space," "receptacle," "nurse," or "mother" of the universe. Taking philosopher John Sallis as my initial guide, I begin with a reading of Plato's *Timaeus*, where the figure of *khora* first appears, and proceed with readings of the cosmological writings of Philo, Origen, Athanasius, Augustine, and the rabbinic commentary *Genesis Rabbah*. Here I am searching for traces of a "forgotten legacy" (as Sallis puts it),[23] one that is both Platonic and biblical yet crucially neither metaphysical nor theological in the usual positivist senses. I am feeling my way toward a "dark cosmology" that can also be conceived as an "eco-chorology." The doctrine of creation ex nihilo lies at the heart of its articulation, somewhat surprisingly. This eco-chorology acknowledges *khora* as the very possibility of possibility; as flow, mutability, unpredictable change; as radical receptivity; as contention, disorder, disintegration; as the ultimate elusiveness of all things—their spooky mystery. It also acknowledges *logos* as *khora*'s necessary double—actualizing, differentiating, initiating, ordering, manifesting. Throughout this first part of the book, the question of whether, how, and why to think theologically—to think the ecological thought by also *thinking god*—recurs.

Part II considers the *Lives* of Saints at the juncture of ecocriticism and queer theory; disability theory and animal studies quickly join the conversation. Taking ecocritic Timothy Morton as my initial guide,[24] I explore how hagiographical literature queers notions of nature or norm, putting the very category of the human into question in part by foregrounding the saint's animality, in part by writing the saint into the landscape. I am also interested in the Saint's *Life* as a kind of performance art that disrupts conventional notions of propriety and beauty, making way for new ones; here the literary *Life* may participate in larger assemblages of texts, objects, and rituals that produce the saint as such. Readings of the *Lives* of Plotinus, Antony, Paul, Mary of Egypt,

Syncletica, and Simeon the Stylite focus on queerings of time, place, and desire and on the performance of the body as both resilient and corruptible, beautiful and broken, but never simply static or whole.

Part III considers material objects from the perspective of thing theory, guided by political theorist Jane Bennett's new materialism.[25] I am interested in the things themselves and in the practices by which ancient Christians cultivate relationships to them as animate beings, at once powerful and vulnerable, protective and in need of protection, lovable and loving. Some of the objects are mobile and relatively small in scale—portable relics and icons, such as Macrina's ring or the bit of the wood of the cross that Paulinus sends his friend Sulpicius. Others take on larger lives as parts of the assemblage of a church or a more extensive built complex, whether in present-day Italy or Jordan or elsewhere. The buildings themselves in turn enter into an often intricate relationship with the surrounding landscape, pointing toward the world as an open mesh of far-flung connection. Here Part III intersects very directly with Part II, as it is in large part the cult of saints that allows the value and agency of things—fragments of wood, metal, clay, blood, oil, wax, stone, bone, glass—to shine forth so intensely on both small and large scales. It intersects as well with Part I, attending not only to things in their barest materiality but also to things invoked in the poetic performance of liturgies of wonder and praise, whether embodied in Basil's hexameral sermons or in Augustine's and Pseudo-Dionysius's cosmic hymns.

The smallest things lead back to the largest things, the largest to the smallest. Viewed through the shifting lenses of an ancient ecopoetics, we humans sometimes loom large—our great aspirations! our great failings! And sometimes we are too small to see at all, absorbed in the dazzle of other lively, active things.

1

Beginning Again with *Khora*

Traces of a Dark Cosmology

Prelude: Anticipations of an Eco-Chorology

> To be sure, the chorology did not become part of the Platonic legacy. . . . The Platonic chorology is rather something held back from the legacy, something not passed along, except for the traces remaining in the text of the *Timaeus*. Yet those traces endure beyond that Platonism that chorology already exceeds; they reach beyond the reach of Platonism as metaphysics. It is, then, from the limit, from a Platonism at the limit of metaphysics, that this forgotten legacy can, as it were, be restored.
>
> —John Sallis

The Platonic *khora*[1] enjoyed a heyday in the last quarter of the twentieth century, drawing the attention of such influential thinkers as Julia Kristeva, Luce Irigaray, and Jacques Derrida,[2] and it has continued to exert its fascination as a figure of semiotic indeterminacy and excess. However, the recent "material turn" is proving less hospitable to this elusive cosmological figure than the "linguistic" or "cultural turn" has been. *Khora*—a Greek word that may be translated as "space" or "place," "land" or "countryside," among other possibilities—lies at the heart of Plato's understanding of materiality, yet it makes virtually no appearance in the writings of the new materialists, object-oriented ontologists, and affect theorists whose work is beginning to shape ecological thinking in provocative ways.[3] For these scholars, Platonism of any kind is more likely to be seen as part of the problem (if relevant at all) than as part of the solution to challenges as vast and as daunting as global warming. Yet my own hunch is that Platonism may harbor a forgotten legacy worth restoring in precisely such a moment as ours.[4]

And so I ask, Where might an explicitly ecological reading of *khora* take us?

Exploring this question will involve returning to the scene of Plato's *Timaeus*. Three friends—Timaeus, Hermocrates, and Critias—gather to host Socrates, laying out a feast of philosophical discourse for their marveling guest. The opening discussion is of politics. Socrates recapitulates the main points of his own discourse on the ideal political constitution, delivered the day before, and Critias promises to provide an embodied example of this ideal in the form of a history of archaic Athens. But first, announces Critias, Timaeus will set the stage by discoursing on "the nature of the universe [τὸ πᾶν], . . . beginning with the genesis of the cosmos and ending with the nature of humanity" (Timaeus 27a).[5] This intruded speech overtakes the entire dialogue; it is performatively excessive, and the difficulty of its topic is repeatedly stressed. Indeed, Timaeus must make more than one start. He begins by marking the distinction between the realms of being and becoming, apprehended by intellect and sense, respectively (28a). Subsequently, he introduces the figure of a divine maker who models the sensory world on an intellectual paradigm, thereby establishing a mimetic relationship between the two realms: what *becomes* is an image of what eternally *is*. However, having played out his narrative of creation within this binary framework, with a little help from the demiurgical deus ex machina, Timaeus is forced to interrupt his own speech. There is a "third kind," in addition to *being* and *becoming*, he confesses belatedly, as he begins his account of "the all" once again (48e). That third kind, as he explains, is "eternal *khora*," apprehended only through "a certain bastard reasoning with the aid of nonsensation, scarcely trustworthy" (52b). This third, it seems, will make all the difference, even if it could not be spoken at the outset. Even if it can hardly be spoken at all.

What is the mysterious *khora* that eludes both sense and intellect yet is somehow (if only just barely) graspable? Some have urged that the word—or, perhaps better, the *name*[6]—is inherently untranslatable, the question of identity inherently unanswerable, with respect to Plato's text. Still, following Timaeus, we may suggest provisionally that *khora* is what provides space or shelter (ἕδρα) for "all things that have a beginning"; it is what allows the invisible forms to shine forth in visible images (52b). *Khora* is something like the nature of nature—so long as we understand that this "nature" is less of a grounding than an abysmal ungrounding and that we ourselves are in the thick of it. "Ecological thought is intrinsically dark, mysterious, and open," proposes ecocritic Timothy Morton, calling for not a green but a "dark ecology."[7] *Khora* seems to respond to that call, as we shall see, opening the way to an apophatic or dark cosmology.

We begin again with the *Timaeus* then, as others before us have. We begin again with a speech that must make a second beginning in order to introduce *khora*—a speech embedded in a text in which "nothing is more vigorously interrogated than the question of beginning," as John Sallis puts it.[8] But we ourselves shall also make a second beginning, considering not only the Platonic dialogue but also its late ancient heirs, in particular, Philo, Origen, Athanasius, Augustine, and the Rabbis who produced the commentary known as *Genesis Rabbah*. We are tracking "Platonic legacies that exceed metaphysics" to borrow Sallis's words again.[9] That is to say, we are tracking disruptive textual moments that point toward other kinds of cosmologies, other kinds of politics. And we are tracking them in unlikely places, including the texts of a theological tradition, both Jewish and Christian, that is generally seen (and not without reason) to have consolidated metaphysics rather than to have disrupted or exceeded it.

Why these five figures? Could we, might we, look elsewhere as well? To be sure. I make no attempt at comprehensiveness, and my choices reflect my own familiarities and proclivities, though they are also neither arbitrary nor surprising, I dare say. I gravitate toward late ancient writers who are not only exegetical in their method but also more philosophical than mythological in their cosmological orientation—assuming that such a distinction holds at all. I also foreground figures and works that can be made to speak to each other in such a way that they tell a kind of story. Finally, I am drawn to texts that are sufficiently innovative and complex as to reward a close reading attuned to what exceeds, hides from, haunts, or even contradicts the main thrust or most obvious sense of the *logos* or argument.

But first, we must turn to the *Timaeus*, where all chorology begins.

Dreaming *Khora*: Plato's *Timaeus*

Khora disrupts and unsettles *logos*. It disrupts and unsettles Timaeus's *logos*, his discourse, forcing him to begin again. *Khora* disrupts discourse as such: although Timaeus assures us that "it must always be called the same thing" due to its unchanging nature (50b), he himself calls it, variously, the receptacle and nurse of becoming (49a), imprint-bearer (50c), mother and receptacle (51a), and finally . . . *khora* (52b). *Khora* (if we can name it at all) also disrupts and unsettles knowledge—specifically, knowledge of being, apprehended through *logos*, or discursive reason. In all of these ways, and crucially, it disrupts and unsettles the distinction between being and becoming. But how exactly does this take place?

With hindsight, we realize that there are already signs of trouble in Timaeus's first speech, even before *khora* makes its appearance. To start with, Timaeus introduces the apparently fundamental distinction between being and becoming as a matter not of truth but of opinion: "Now in my opinion [κατ' ἐμὴν δόξαν] we must first distinguish the following," he begins (27d). We must first distinguish, that is, between what is apprehended by *nous*, or intellect, with the aid of *logos* and what is apprehended by *doxa*, or opinion, with the aid of sensory perception that lacks *logos*. At least, that is Timaeus's opinion; that is how things seem to him. What are we to make of such a surprising reluctance to claim the authority of truth for this much-vaunted dualism? How can mere opinion secure the objects of intellection? The foundation of Platonic metaphysics—the privileging of being over becoming—seems a bit shaky already. As Sallis argues, with respect to this passage, "Rather than simply reasserting an established distinction, the *Timaeus* reopens the *question* of the distinction. Thus, a certain suspension is operative here at what seems to be the beginning of Timaeus' speech, here where he seems to make a beginning, to begin with the beginning."[10]

Equally striking is the fact that Timaeus immediately has recourse not to reason but to necessity, as he continues his explication of the genesis of the cosmos: "All that comes to be must of necessity come to be from some cause" (28a).[11] Later, he will introduce the discourse on *khora* as an attempt to lay out "what comes to be through necessity," distinguishing this from the first discourse, which is ostensibly concerned with "what is crafted [δεδημιουργημένα] through intellect," not necessity (47e). Yet the first discourse is also entangled with necessity from the start. At Socrates's prompting, Timaeus acknowledges that necessity compels him to invoke the gods and goddesses before beginning his speech (though it is not clear that he actually does so) (27c). And now it is necessity—specifically, the need for a cause for becoming—that compels Timaeus to introduce a god, himself a divine artisan (δημιουργός) who will craft the universe in imitation of intelligible forms (28a). Both like and unlike *khora*, this god, invoked out of necessity, goes by more than one name—artisan, framer, maker, constructor, even father—and proves difficult to comprehend. In relation to being and becoming, is he not himself a "third kind"? Where does he come from? Timaeus dodges such questions: "Now to discover [εὑρεῖν] the maker and father of this universe [ποιητὴν καὶ πατέρα τοῦδε τοῦ παντὸς] is hard work, and it is impossible for the one who has discovered [εὑρόντα] him to speak to all" (28c). The verb here translated as "discover" can also mean "devise" or "invent." The maker is a fabricator but may also be a fabrication himself then[12]—a kind of necessary fiction that grants the account a beginning it would not otherwise have. And yet, as beginning and cause, the artisan god withdraws from both knowledge and speech, and Timaeus will ultimately have to discover a new beginning. (It too will withdraw.) The suspicion that the initial narrative of divine creation is tinged with fiction is only intensified when Timaeus immediately reminds Socrates that precision and certainty will prove elusive with regard to such matters as "the gods and the genesis of the universe"; Socrates will, therefore, have to be content with a likeness of what is already a likeness—that is to say, a "likely story" (ἐικός μῦθος) (29d).[13] *Doxa,* or opinion, in other words.

As the account of the genesis of the universe unfolds in more detail, the model of creation as a *techne,* or craft, dominates, secured by the figure of the fabricating god. Timaeus thus attempts to place *poiesis* firmly on the side of culture as opposed to nature, despite his frequent use of the term *genesis*, with its connotations of erotic generation. The cosmos is crafted by a maker and comes to be according to a plan, imitating the perfection of an invisible form or idea.

But what is that form, that idea, that invisible vision? It is not goodness or beauty, as readers of other Platonic texts might expect. It is an invisible animal (ζῷον). The cosmos is a visible animal, blended of body, soul, and intellect, and it is modeled on an invisible animal. Like its paradigm, it encompasses all other animals: it is an animal filled with animals (30c–d). A perfect spinning sphere, complete and contained, it has no outside: there is nothing outside it. Indeed, it eats its own excrement (33d). After all, it is not a mere planet, as our own cultural imaginary might falsely suggest. It is the universe—*to pan*, the all—and thus it must include everything that is. And everything, it seems, is an animal, a *zōon*, a living thing. But everything is also a god, a blessed god (34b)—an animal-god, a *divinanimality*,[14] that eats its own excrement. Even more surprisingly, perhaps, we learn that the universe is not strictly generated stuff but is a complex mixture of being, becoming, and, a third thing, soul, itself a mixture of being and becoming (35a). Mimesis, which at first seems to distinguish being from becoming, exemplar from copy, now appears to enfold being within becoming quite cunningly. Cosmos as the product of *techne* is thus already destabilized in this first discourse: culture and nature are not altogether distinct where the cause is enfolded in what is generated, and what is generated is called an animal. Tellingly, when

Timaeus describes the god's joy in perceiving his creation "moving and alive," he refers to the god not as maker but as father (37c).[15]

The motion of the cosmic animal, as Timaeus goes on to describe it at some length, is manifest above all in the ordered and majestic wheeling of the heavens, the "choral dance" of the celestial bodies (40c). *Ouranos,* or sky, is established as "the eternal image of eternity, moving according to number—what we have named time," notes Timaeus (37d). This is "uranic time," as Sallis calls it,[16] a time not withdrawn into the human soul but proper to the soul of the cosmos, shared by the animals it contains. Yet there is a hint that still other temporalities might also be imaginable—a time of the earth, for example. "Earth our nurse . . . was constructed to be the guardian and creator of night and day, being the first and oldest of the gods," Timaeus remarks, somewhat perplexingly, at the end of his discussion of the heavens (40b–c). Is earth older than sky after all?[17] Does it too have a time? Perhaps a time that moves unimaginably slowly, if indeed it can be said to move at all. Considering a stone shining in the sunlight, Sallis remarks, "One will sense that the time borne by the stone is not a time measured by the movement of the heavenly bodies but rather a time to which movement is absolutely alien, a time of utter repose, a time that is utter repose. A time withdrawn even more decisively from that of human events, a time that is neither of the sky nor of history. A lithic time, perhaps. A time of the earth."[18] Is earth time—geological time—also khoric time? Or would khoric time not constitute an "undulating temporality," arising in the interplay of entities existing within vastly different modes and scales of time, as Morton frames it?[19]

With this, let us now be drawn again into the second discourse, the discourse on *khora* proper—though of course there is nothing proper about *khora.* Timaeus opens by confessing that the universe actually has not one cause but two. In addition to *nous,* or intellect, which is at work in the god's imitation of the intelligible forms, there is another cause, namely, *anagke,* or necessity. Necessity is described as an "errant cause" (48a): although open to persuasion, so to speak, it wanders and strays; it resists and eludes the full control of intellect or forms. In acknowledgment of this formerly unnamed, even suppressed, cause, Timaeus must make a fresh start, beginning again, as he announces more than once (48b, e). Having introduced a second cause for genesis, he must now also introduce "another, third kind" (48e). Model and copy are no longer enough; the discourse needs a third, which is, however, "difficult and obscure." Timaeus's rhetoric drags with reluctance as he is caught between a rock and a hard place—the necessity to speak of the third and the near impossibility of doing so. But he presses on. "What power are we to suppose it to have by nature? This especially, that it is the receptacle, the nurse, as it were, of all genesis" (49a).

Having initiated the discussion of the third, Timaeus is once again forced to back up, declaring that there is more to be said but only after he has first discussed "fire and those with fire" (49b)—that is, water, air, and earth—as they were before the genesis of the universe. However, these would-be elements prove to be little more than fugitive figments, impossible to pin down. Each is constantly in a state of transformation into the other; in that sense, none can really be said to exist. Each flees any naming of "this and that" that might suggest stability (49e). In the beginning, it is transience all the way down, so to speak, and one can say no more than that something appears fleetingly "suchlike" (firelike, waterlike, etc.) (49d–e).[20] To cite Sallis again, "rather than being captured by λόγος and assimilated to it, fire and the others flee from it, retreat before it, elude it."[21]

At this point, Timaeus circles back to the third kind, or *khora*, albeit without naming it as such. How in this context can the terms "this and that" be safely applied? he asks. Not to the fleeting apparitions of fire, water, and so on, but only to that "in which" each appears in being generated and "from which" each perishes again (49e–50a). That which "receives all bodies" remains unchanged in its very formlessness: "It is always receiving all things, yet nowhere and in no way does it ever take any form like any of those entering it" (50b–c; cf. 50d–e). This stability evidently enables it to be named, however provisionally. Nonetheless, like wax that is imprinted, what makes it stable is, paradoxically, its perpetual instability: "Being moved and shaped by all that enter it, through them it appears different at different times." Such imprinting is none other than the process of mimesis by which the eternal forms are copied, we now learn, "in a manner that is hard to declare and wondrous, which we shall consider hereafter" (50c). Yet again Timaeus defers a full discussion, offering a provisional summing up: "We must conceive of three kinds, that which becomes, that in which it becomes, and that from which what becomes, having been copied, is begotten." Shuffling the order, he compares the three to a mother, a father, and their offspring (50c–d). Contrasting demiurgical and erotic models of production are once again juxtaposed, as if the binary must continually be both spoken and unspoken. At the same time, Timaeus has, almost by sleight of hand, made a transition within his discourse: having begun with the fugitive elemental flux that is sheltered by, and very nearly identified with, the receptacle, now he speaks of the receptacle as the site of generation of copies of forms that can be named "this and that." He returns to the question of the naming of the receptacle itself. The "mother and receptacle" of generation should not be called earth, air, fire, or water, Timaeus emphasizes; rather, "we shall not lie if we call it a form that is invisible and unshaped, all receiving, partaking of the intelligible in some most perplexing and baffling way" (51a–b).

At this point, Timaeus once again seems on the verge of unraveling his own discourse. Is there really anything beyond what we can perceive with our senses? he asks. Are the intelligible forms of things no more than the product of *logos* or discourse (51c)? Much as he previously expresses his opinion, Timaeus now declares that he casts his vote for the view that the forms do indeed exist. (Is the existence of forms to be settled through the vagaries of a democratic process, of which Plato was famously suspicious?) Once again he enumerates the two kinds of knowing and their corresponding objects, form and image. The form is perceptible to the intellect; it is self-identical and unchanging. The image is perceptible to the senses; it is mobile and transient. But it also has the same name as the form and is similar to it (52a). Here Timaeus distinguishes quite sharply between being and becoming, even placing them in opposition, and also aligns the two more closely than ever, identifying them as virtually one. That which is generated is called by the same name as its eternal form; it resembles the form. However, if intelligible form is by definition self-identical and self-contained, neither receiving nor being received, as Timaeus also stresses, how can it be copied or duplicated? How can something else be called by its name? In the first discourse, Timaeus introduces the god to circumvent the problem; here he does not.

It is "the question of the doubling of being," as Sallis puts it, that brings us to the brink of what he calls the *chorology*, "in which Timaeus' discourse comes as near as it ever will to the beginning."[22] Once again, the "third kind" is invoked, supplementing the first two; now it will receive its name. Let me cite the entire passage, following Sallis's translation, which clings as closely as possible to the nearly impenetrable Greek:

> Moreover, a third kind is that of the χώρα, everlasting, not admitting destruction, granting an abode to all things having generation, itself to be apprehended with nonsensation, by a sort of bastard reckoning [λογισμῳ τινὶ νοθῳ], hardly trustworthy; and looking toward which we dream and affirm that it is necessary that all that is be somewhere in some place [τόπος] and occupy some χώρα; and that that which is

> neither on earth nor anywhere in the heaven is nothing. As for all of these and others akin to them and concerning [their] wakeful and truly underlying nature, under the influence of this dreaming, we are unable to awaken, to distinguish [these], and to say the truth: that for an image, since not even that itself on the basis of which it comes to be generated belongs to the image but it is always brought forth as the phantom of something other—because of this it is appropriate for it to be generated in something other, clinging to being at least in a certain way, on pain of being nothing at all; whereas to the aid of that which *is* in the manner appropriate to being there comes the precise true λόγος: that as long as one thing is something and another is something else, neither of the two will ever come to be in the other, so as to become at once, one (and the same) and two. (52a–d)[23]

Khora eludes both intellect and senses, as we have noted before. It can only be intuited through a kind of knowing that is alien and transgressive with respect to intellect or reason, Timaeus explains—a "bastard reckoning." Put otherwise, it can only be dreamed. Or rather, it can only be apprehended by one who both dreams and awakes from the dream. In the crossing of dreaming and wakefulness, the transience, mystery, and phantomlike nature of things is revealed; one may perceive the dream itself as a dwelling place for otherwise rootless images. It is only a small leap to suggest that Timaeus here intimates that *khora* is the dream—perhaps better, the dreaming—of the universe. As such, it sustains the constant flux of existence: χωρέω can be translated as "flow." It also enables *things* to emerge within that flux—leaf, human, frog, star ("one thing is something and another is something else"). That is, it enables things to appear to one another, to sense and be sensed, to affect and be affected, in not just a doubling but a relational multiplication of being: a leaf presents itself differently to a human, a frog, or a star. We perceive the sensual images; things-in-themselves may be partly known but can never be fully grasped. That much wakefulness is impossible, Timaeus suggests (astonishingly, from a certain "Platonic" perspective); it only *seems* possible when the third is suppressed. Things-in-themselves—the embodied forms—withdraw from us; all things withdraw from one another. And that too *khora* enables: χωρέω can also be translated as "give way or withdraw."[24]

As Morton puts it (echoing the phenomenological position adopted by recent object-oriented ontologists), there is always a "dark side" to things, a concealment that accompanies their very revelation in and as sensual images;

"things are themselves but we can't point to them directly."[25] Of course, object-oriented ontologists would understandably resist conflating Plato's forms with their objects or things. Graham Harman draws the distinction sharply: "The objects of object-oriented philosophy are mortal, ever-changing, built from swarms of subcomponents, and accessible only through oblique allusion."[26] Yet the gap may not be so great for the reader of the *Timaeus*. After all, from the perspective of *khora*, what are forms? Are they even real? We can say this much: forms secure the existence of objects per se without reducing them to discourse; forms protect difference, within (and beyond) the flow of existence, within (and beyond) the embeddedness of things in webs of relationality.[27] Forms are immortal in some sense, yet forms need *khora* (necessity is yet another name for the third kind), and *khora* is also immortal—unchanging precisely insofar as it is ever-changing, swarming, and "accessible only through oblique allusion."

Pausing for a moment, we might ask, If the chorology at the heart of Timaeus's second discourse "comes as near as it ever will to the beginning," why does Timaeus not simply start here? Why does he have to begin again? The answer seems to be at least twofold. As we have seen, *khora* cannot be approached directly: it is only through the first discourse on being and becoming, form and image, that the need for a third kind becomes evident, in the tacit posing of what Sallis names "the question of the doubling of being." It is also the case that *khora* resists linear chronologies. This beginning is not an origin, and it presses toward no end; rather, it opens time, unfolding and enfolding dissonant and looping temporalities. Constantly in motion, it is always in the middle—in the midst of becoming.

When Timaeus resumes a more linear narrative that attempts, for a brief interval, to accommodate *khora*, the effect is comical if also instructive. Before the universe comes to be, there are three kinds, being, *khora*, and becoming, he recites. *Khora*, once again feminized and dubbed the "nurse of becoming," is in an agitated state: "being liquefied and ignified and receiving the shapes of earth and air as well, and suffering all the other affects [πάθη . . . πάσχουσαν] that accompany these, she manifests appearances of all sorts." The nurse is not only radically unstable in her appearance but is also powerfully buffeted by the disordered potencies with which she is filled: unbalanced, swaying wildly, "she is herself shaken by these and, in turn, being moved, shakes them" (52d–e). Agency and affectivity are here thoroughly decentralized: *khora* suffers the sufferings, is affected by the affects, of all of the fleeting and chaotically dispersed elemental forces that appear in her, fill her, shake her. Yet her seemingly abysmal passivity is also a kind of agency, for when the potencies move her, her own motion moves them too. Moreover, the collective agency emergent in *khora* develops focus and force, and the shaking—now attributed to "the one receiving" (i.e., *khora*)—has the effect of winnowing and sorting the "four kinds": water, fire, earth, and air (53a). Still, they are only traces of themselves (53b), not yet "this and that."

Now the resumed narrative takes a decisive turn, bending back toward its own first beginning: exit *khora*, and reenter the creator god, who has been absent for quite some time, as Timaeus notes (53b). Indeed, *khora* and the god seemingly cannot occupy the same discursive space. They embody incompatible models of agency and generativity, nor is it possible to speak of them in the same way. For the one, there is a "bastard reckoning," for the other a "likely discourse" or "story"; if the former is characterized by extreme difficulty, bringing discourse very nearly to a halt, the latter may be suspected of frivolity, undertaken more for pleasure than for profit, as Timaeus acknowledges (59c–d). Opting for the pleasure of narration, Timaeus goes on to tell how the artisan god makes proper elements out of the khoric traces, constructing them according to geometrical principles; from these elemental bodies arise many (indeed, countless) other kinds of bodies, each characterized by distinctive affective properties (παθήματα). At this point, Timaeus has arguably both overwhelmed himself and caught up with himself, so to speak. A third discourse can (and perhaps must) be commenced, picking up where the first one left off (69a–b). It is devoted almost exclusively to the creation of the mortal soul and body of humans.

It should be noted that Timaeas has already dealt with the genesis of humans in his first discourse before interrupting himself and making a second beginning. That first account includes at least one notably khoric moment, when the spherical bodies that initially house the immortal souls roll around wildly, banging into each other and spewing excrement in all directions (43a–c). However, it focuses primarily on the immortal soul, and its discourse of the body is oriented toward and governed by the head, the body's divine ruler (44d). Post-*khora*, the human body can be spoken differently, it seems, even if *khora* itself cannot be assimilated into the "likely account" with its demiurgical divinity. The third speech offers a discourse of the body below the neck—the body of a living being or animal, teeming with other animals, as it were. Their coexistence is not without contentiousness.

According to Timaeus, the human soul is a triad of mind, spirit, and appetite, and each has its seat in a particular region of the physical body. The spirited soul is confined to the area between the neck and the midriff, where it can take instruction from the mind, a divine *daimon* located in the head, while also keeping the appetitive spirit under control; heart and lungs play a role in regulating the spirited soul's level of excitement as well, each also exercising a kind of agency (70a–e). The appetitive soul is "bound" between the midriff and the navel "like a savage creature" (ὡς θρέμμα ἄγριον), dangerous but necessary to the body's well-being (70e). Its guardian organ is the mirror-like liver, "which receives impressions [from the mind] and provides visible images" (71b) that either frighten or calm the savage creature. Intriguingly, these images curb the excessive desires of the appetitive soul by directing it toward another kind of knowing (a "bastard reckoning" perhaps?), manifest as the prophetic insight that comes in divinely inspired dreams or visions, when reason sleeps (71e). No surprise that such tensive inner balances cannot always be maintained in this notably unstable human assemblage. Body and soul—the hybrid *zōon,* or animal—may become diseased. And the best remedy for disease, explains Timaeus, is "if one imitates the one that we have called the nurturer and nurse of the universe" by keeping the body in constant motion (88d). As with *khora*'s winnowing dance, internal vibrations can shake internal parts into their proper place. Thus the cure comes from the same source as the disease, so to speak.

Have the god and his helpers lost control of the process of intelligent design, as Timaeus tells it? Correspondingly, is intellect betraying its limits? Neither the cosmos itself, human animals, nor their political collectivities (πολιτεῖαι) (87b) are simply governed by reason, Timaeus seems to suggest. The mind needs the liver and gut not only to sustain life but also to open it to divine insight. Forms need *khora* in order to enter into the play of motion and affectivity inherent to bodies of all kinds. But *khora* does not only sustain movement and interaction, as if these were in tension with or even opposition to stability and individuation. Ultimately, the emergence of things as things also rests on the dispersed yet coalescing movements and intensities associated with *khora* in this text. That the god makes copies of the forms is a pleasant fiction, offering a semblance of truth. The tortuous, bastard reckoning by which one approaches *khora*, as if dreaming, does not even offer a semblance of truth, because *khora* does not resemble anything. And yet something lodges itself in discourse, if only through repeated comedic refractions: the phantom figure of the nurse of genesis, reeling and swaying, jiggled and jiggling.

Timaeus's speech, and with it the dialogue itself, ends on a distinctly khoric note, in the performance of what Sallis aptly dubs a "descensional bestiary."[28] This final section begins with Timaeus's proposed explanation of human sexual difference: in short, cowardly men are reincarnated as women. Devolution thus gives rise to multiplicity, which in turn leads to desire. Once again, Timaeus describes a condition of bodily imbalance and agitation that tacitly recalls *khora*'s state before creation. The male genital organ is "disobedient and self-willed, like an animal resistant to *logos*" (91b); the female womb is likewise an "animal" that "wanders all through the body" causing great distress (91c). The remedy for this disease? Generative shaking, yet again—sexual intercourse, that is—brought on by the very "desire and love" that created the disorder to start with. Through this joining, invisible and unformed lives are planted in the womb, there becoming differentiated into distinct creatures who are nourished and brought to the light—"and thus they complete the genesis of living creatures [ζώων]" (91d).

The generative process of human becoming-other does not end with men being transformed into women, however. Sexual difference is just the beginning of animal difference. Other men are transformed into birds, still others into four-footed beasts, wriggling reptiles, and brainless fish. "And in all these ways all living creatures (ζῶα), then and now, pass into one another," Timaeus sums up (92b–c). The descent is also an ascent, it seems, and perhaps it is no longer easy to determine what is up and what is down. Animals—and for Timaeus, all things or beings are animals—are constantly turning into each other, much like the traces of fire, air, water, and earth are "always becoming something else" (49d) and "are always circling round" (49e) in *khora*. Having arrived at this point, Timaeus announces that his "discourse on the all"—really three discourses—has now finally reached its *telos*, or goal (91c).

It is a striking place to stop. But is this stopping point really an ending? This discourse will receive a supplement, albeit a fragmentary one—the incomplete dialogue *Critias*. As we recall, Socrates presents his companions with a depiction of the ideal *politeia*, or political constitution, on the day before our dialogue takes place. Dissatisfied with the static character of his own depiction, Socrates then expresses his desire (ἐπιθυμία) to see the *politeia* in motion (19b)—specifically, to see it in a state of war (19c). It is in response to Socrates's desire that Timaeus first sets the stage with his chorological account of the beginnings of the universe and of humanity. Now Critias takes it upon himself to fill out the narrative he has hitherto only briefly sketched—namely, the history of Athens's fateful war with Atlantis, said to have taken place nine thousand years previously.

If Socrates gives us an ideal *politeia*, a city that is nowhere, Critias gives us a *politeia* that is crucially emplaced—that exists in a specific *khora*. A khoric city,[29] its existence both enabled and limited by the particularities of geography. In the *Timaeus*, Critias cites the words supposedly uttered by an Egyptian priest to the legendary Athenian statesman and lawmaker Solon: the Athenian *khora* gives birth to "the most beautiful and best race of humans," as if from a seed implanted in a womb (23b). The excellence of the Athenians is a reflection of the excellence of their land. Yet the same priest also relates that the Athenian land has a tendency to flood, washing away its cities along with their archives, thus leaving its people forgetful of their own history (22e).

Athenian memory erodes, then. In particular, the memory of the island of Atlantis has washed away—Atlantis, which long ago went to war to enslave the Mediterranean region only to be heroically defeated by the Greeks. Now the imperialistic city has been swallowed by oblivion, buried beneath the sea (25b–c). Forgetfulness of this history makes the Athenians childlike, as the priest declares (22b), but it does not make them innocent. Forgetful Athens has been lured by imperialistic ambition. To this the presence of the Syracusan general Hermocrates silently attests. As Sallis notes, "To an Athenian of the mid-fourth century he would be recognized above all as one of those most responsible for repulsing the greatest effort Athens had ever made at imperialist expansion."[30] The inclusion of Hermocrates in the gathering thus reminds us of what the Athenians have forgotten. The needs and desires of the city—not only for bodily nourishment but also for materials for the crafts (δημιουργίαι) (Critias 110c)—are fed by the countryside, by the land; and overweening desires will lead the citizens to go to war in order to capture more land.[31] The khoric city is subject to the stirrings of both desire and aggression.

In the fragmentary supplement, Critias dwells at length on the virtues of the Athenian *khora*, which in his own time still rivals all others in agricultural and pastoral production, he claims. But he dwells at even greater length on the depletion of the land: its bounty is nothing in comparison to what it once was. Millennia of erosion "have left only the wasted body of the *khora*," much like "the skeleton of a diseased body" (111b). The forests have been felled, the thin soil no longer holds moisture; only shrines remain as markers of the springs that once flowed (111c–d). Critias blames natural disasters (111b). Yet war also ravages the land, and so too does overproduction, and both of these factors are also in play in his discourse. Memory of the dangers of desire and war wash away, and so too does the earth that nourishes life.

We cannot simply conflate the mysteriously cosmological *khora* of Timaeus's speech with the geopolitical *khora* of Critias's speech. Yet we also cannot ignore their convergences. As Sallis observes, a strong analogy is established: what *khora* as third kind is to the binary of the intelligible and the sensible, *khora* as earth or land is to the binary of ideal and actual cities.[32] From the start, Timaeus's chorology is implicated in a political problematic. Cities are not contained or autonomous, as Socrates's account may have seemed to imply, and political constitutions do not merely involve humans. Rather, khoric politics, like khoric cosmology, demands an attempt to think with and as the totality—*to pan*, the all, the animal teeming with animals. It calls for something like what political theorist Jane Bennett names a "vital materialism," enabling a shift from an anthropocentric environmentalism to an inclusive "political ecology of things." "If I live not as a human subject who confronts natural and cultural objects but as one of many conative actants swarming and competing with each other, then frugality is too simple a maxim," writes Bennett.[33] What she names abstractly a "conative actant" is something like a *zōon*—an animal, living being, object, or thing that can be called "this and that," that possesses some kind of agency. It is not enough for humans to be frugal or live modestly, to make way for "nature." Humans are part of the khoric swarm.

Timaeus declares that his speech has reached its *telos*, but *telos* cannot here mean closure: the chorology that erupts in the middle of the discourse inscribes a radical openness at the heart of cosmology. Critias, for his part, delivers a speech that is doubly incomplete and thus also doubly open: the first attempt is anticipatory and truncated, calling for a supplement, while the supplement itself arrives as a broken or unfinished fragment. Socrates's imaginary city was never meant to be the last word. We are just beginning, we are always just beginning, to think *khora* and its politics.

Interlude: Fragments of an Eco-Chorology

Beginnings are glimpsed in medias res—as we leap into the middle of things.

Causality weds intention to necessity.

Intention and necessity pervade the universe.

The figure of the artisan/father god is a mere image of the truth—a fiction or "likely story."

The artisan/father god may be a necessary fiction, if we want an account that begins at the beginning and ends at the end.

Do we want such an account?

It is possible to say: the universe is a *zōon*, an animal, a living being—an animal filled with animals.

All things are animals or living beings. There is not life and non-life.

There is not animal, vegetable, and mineral.

There is not human and non-human, subject and object.

There is not culture and nature.

It is possible to say: the universe is a god.

There is nothing and no place outside the universe.

Not even god?

To give an account of the universe without the fiction of an artisan/father god is most difficult.

It entails driving discursive reasoning to, and then beyond, its limits.

It entails driving sensation to, and then beyond, its limits.

Being is enfolded in becoming.

The universe flows; it is its many flowings; all things are constantly becoming other.

The universe is an animal-thing, filled with animal-things that are in turn filled with animal-things.

The universe is constituted by difference, desire, cooperation, and contention; also, by indifference.

Things manifest themselves to others.

Things withdraw from others; they harbor mystery.

There is not active and passive.

Human communities go to war for land.

Human communities waste the land; they lay waste to the land; they create wastelands.

Ultimately, the universe must consume its own wastes.

Humans forget this.

The earth has wasted away; it is a sick body.

Politics is not just about humans.

Humans forget this.

Endings open onto new beginnings.

Is it time to begin again?

Khoric Legacies: Readers of *Timaeus* and Genesis

Interpretation of the *Timaeus* was contested from the start. Aristotle complains that Plato's concept of the third kind is both unclear and superfluous: "What he has written in the *Timaeus* is not well defined. For he has not stated clearly whether the all-receiving one [πανδεχές] is distinct [χωρίζεται] from the 'elements.' Nor does he make any use of it, saying that it is some substratum [ὑποκείμονόν] prior to what are called 'elements,' as gold is to things made of gold." Having made Plato say what he does not in fact say (that *khora* is "some substratum"), Aristotle goes on to conflate the "nurse" of the *Timaeus* with "primary matter [τὴν ὕλην τὴν πρώτην], which is not separate [χωριστὴν] but always with its opposite"—that is, form (On Generation 329a). This reduction of *khora* to the substratum or matter of formed bodies is momentous and has far-reaching effects, not least in rendering *khora*, in the guise of *hyle,* or matter, at once passive and (very nearly) knowable. *Khora* is now read as "stuff."

The major target of Aristotle's critique of the *Timaeus,* is, however, its apparent claim that "the world, although indestructible, was generated" rather than eternally existent. He notes that some of Plato's followers attempt to defend their master by asserting that the narration of the world's beginning is not meant to be taken literally. Rather, it should be understood as a pedagogical device: "They say that they speak of generation, not meaning that the world was ever generated, but for the sake of teaching" (On the Heavens

279b–280a). Aristotle rejects this interpretation of the text, even if he agrees with Plato's defenders on the point of doctrine: the cosmos has no point of beginning. This remains the position of the Platonic Academy. As John Dillon puts it, "The first principle on which all Old Academicians seem to agree, closing ranks against the tendentious interpretation of Aristotle, is that the central myth of the *Timaeus* is not to be taken literally."[34] Later followers of Aristotle are inclined to agree with their Platonic colleagues in privileging a nonliteral reading of the account of cosmic genesis.[35]

The *Timaeus* was not, then, ignored in the centuries immediately following its composition. However, it acquired a new level of cultural cachet in the Roman imperial period. As Heinrich Dörrie writes, "Suddenly the *Timaeus* was in every mouth; as surely as every Greek knew his Homer, so surely did every educated person know their *Timaeus*."[36] In the wake of Antiochus of Ascalon's rejection of Academic Skepticism, the text, now interpreted more literally, played a central role in the emergence of a dogmatic and eclectic brand of philosophy that modern scholars commonly refer to as Middle Platonism. Alexandria in particular nurtured a milieu in which Jewish, Christian, and pagan intellectuals participated in a shared philosophical discourse. As Carl O'Brien has recently argued, "They [were] all attempting to various extents to respond to the nature of demiurgy, as advocated in the *Timaeus*."[37] A common text created opportunity for differentiation as much as for identification or connection: by the second century of the common era, the *Timaeus* had become a site of contestation between Christians and pagans, as philosophers like Celsus lauded the authority of the Platonic texts while criticizing the literalizing interpretations of their Christian counterparts, much as Aristotle had once criticized Plato (whether fairly or not). Maren Niehoff suggests, "More than other Platonic works [the *Timaeus*] helped to construct a textual community, which sought to preserve the original Greek tradition against its appropriation by Christian readers."[38]

The popularity and disputed interpretation of the *Timaeus* provide crucial historical context for our pursuit of "khoric legacies." However, the history of reception of the Platonic text is not our primary concern. Rather, we seek "the archaic or exorbitant moments that lie, for the most part concealed," within certain Jewish and Christian cosmological texts of late antiquity.[39] While Philo, Origen, and Athanasius of Alexandria surely knew the *Timaeus*, as did Augustine via Cicero's translation, there is no clear indication whether the midrashists of *Genesis Rabbah* were familiar with it directly. But this is of no great import. As readers of the biblical book of Genesis, all of these thinkers participate in a late ancient discourse on the nature and origins of the cosmos—a discourse that is heterogeneous, contested, and complexly intertextual but by the same token also, in some significant sense, shared. Rather than merely asking how they have interpreted the *Timaeus*, we shall reopen a dialogue between their texts and the *Timaeus*. In so doing, we may hope to give voice to eco-chorologies at once ancient and new.

Philo of Alexandria, *On the Creation of the World*

Niehoff dubs the Alexandrian-Jewish philosopher Philo of Alexandria (c. 20 BCE–c. 50 CE) "the first Platonist"—the first, that is, to read Plato's texts as "a source of ancient authority." Identifying Plato as a "most holy" philosopher, "Philo extends the authority of the Scripture to Plato's works," she suggests. In this respect, Philo was a harbinger of things to come.[40] By the beginning of the second century, the Platonic corpus in general and the *Timaeus* in particular were widely regarded as possessing a quasi-canonical status, as we have noted. Thus, Philo was on the forefront of the development of Middle Platonic thought, a significant contributor to the rich exegetical and philosophical culture of Alexandria.

As David Runia demonstrates in a monumental study, "The *Timaeus* casts a long shadow over the writings of Philo."[41] Philo is particularly fascinated by the resonances between Platonic and biblical cosmogonies.[42] Here we shall focus on his treatise *On the Creation of the World*. As we shall see, this work offers an exposition of the biblical account of creation that proceeds from the assumption that Genesis and the *Timaeus* speak with one divine voice. Philo follows Plato in giving cosmology a political framing, although he differs in emphasizing the universality of the Mosaic *nomos*, or law: "The world is in tune with the law and the law with the world, and the law-abiding man is thereby a world-citizen [κοσμοπολίτου]" (On the Creation 3).[43] In many other respects too, Philo's reading of Genesis converges with the *Timaeus*. Yet the figure of *khora* is largely suppressed: indeed, at first glance, *On the Creation* seems to offer not a chorology but an antichorology. For Philo, the divine "maker and father" (On the Creation 7; cf. Timaeus 28c)[44] is the one and only cause of the universe. The god creates the intelligible form of the world as well as its sensible copy (On the Creation 19). Moreover, unlike Timaeus, Philo does not begin again. The demiurgical narrative remains seamless. There is no hint that the god might be merely a necessary fiction or that anything might be lacking from the initial account.

This is not to say that *khora* is altogether absent. By merging and simplifying what in the *Timaeus* are two distinct and competing discourses, Philo brings *khora* inside the demiurgical narrative. (Whether it enters as subdued captive or Trojan horse remains to be seen.) The following passage is instructive with respect to his method: "If one wished to search for the cause on account of which this universe was crafted, it seems to me that one would not err in saying, as one of the ancients did say, that the father and maker was good, for which reason he did not grudge the substance [οὐσία] a share of his own excellent nature, since it had nothing good of itself, but was able to become everything" (On the Creation 21). As Runia points out, Philo here conflates two passages from the *Timaeus.* One famously refers to the god as "maker and father" (Timaeus 28c). Another asserts that "he was good, and in the good no envy concerning anything ever arises; being without this, he wished that all things [πάντα] would come to be as like him as possible" (Timaeus 29e).[45] Philo substitutes language of *ousia,* or substance, for Plato's "all things." Yet this is not substance in the Aristotelian sense of formed matter.[46] According to Philo it has "nothing good of itself," being "able to become everything," much as Plato's *khora* has no form of its own but is able to receive all forms. Thus, by incorporating *khora* into the demiurgical narrative, Philo not only follows Aristotle in reducing it to passive matter (what Aristotle would call *hyle*); he also elides the distinction between *khora* and the realm of generation ("all things"), restoring a simple binary of being and becoming, noetic model and sensible copy. Philo further innovates by identifying the noetic model with divine *logos,* or reason—the god's intention or thought, as it were (On the Creation 24). Thus, the binary can also be reformulated as god and world or creator and creation. There would seem to be no need at all for *khora* as a third kind.

Traces of khoric disruption remain, nonetheless. Philo has opened with an invocation of Aristotelian hylomorphism, assuring his readers that "in all existing things there must be an active cause and a passive object," the former identified with the "intellect of the universe," the latter with "something without life and motion in itself, which, when it is moved and formed and enlivened by intellect, changes into that most perfect work, this world" (8–9). Matter is inert and lifeless stuff, then. Now, however, he suggests that matter or substance is "full of difference, discord, and incoherence"; the deity must master the realm of generation, overcoming disorder with order (22). Which is it? Is substance passive or is it active and chaotic? As Runia points out, Philo here seems to be influenced by a passage in the *Timaeus* that occurs just after the one he has cited in the previous section: "When the god took hold of the visible universe, which was not at rest but moving discordantly and irregularly, he brought it into order out of disorder" (Timaeus 30a). Nonetheless, "the discrepancy is a puzzle," as Runia also notes. "Where does the disorderliness come from?"[47] The Platonic intertext invites a further question: if Philo's substance is active as well as receptive, is its movement in any sense positively generative—as suggested by Timaeus's comparison of *khora* to a swaying, quaking winnowing basket? Or is it sheerly resistant and negative, a force to be overcome?

Philo seems ambivalent yet inclined toward the latter option. He goes on to state that the god's powers to confer gifts are vastly greater than the capacity of generated things to receive them. Indeed, had the god not restrained the greatness of his powers, the latter "would have sunk under them [ἀπεῖπεν]" (On the Creation 23). Does the image hint at the possibility of violence? As Runia observes, "Philo's emphasis on the superiority of the divine power is extremely graphic."[48] Paradoxically, the very stress placed on the god's potency leads to acknowledgment of its limits. Even passive matter resists divine will, if only in its weakness and finitude. Perhaps tellingly, Philo remains unclear about whether matter is uncreated or divinely generated.[49]

Other ancient readers of Genesis 1 would discover primordial matter, whether created or uncreated, in its second verse ("the earth was invisible and unformed"), but this is not Philo's reading.[50] It cannot be, because he understands the account of "day one" of creation in the first chapter of Genesis to refer to the noetic world on which the sensible world is modeled. He begins his exegesis with the first three verses: "In the beginning the god made the heaven and the earth. And the earth was invisible and formless [ἀορατος καὶ ἀκατασκεύαστος], and darkness was above the deep [σκότος ἦν ἐπάνω τῆς ἀβύσσου]; and a divine spirit-breath was moving above the water. And the god said, 'Let light come into being'; and light came into being" (Gen. 1:1–3, LXX). Philo discovers in these lines the divine creation of seven noetic forms. "Now first the maker made an incorporeal heaven and an invisible earth and the form of air and of empty space. . . . Then [he made] the incorporeal substance of water and of spirit-breath and, to crown all, light" (29). Repeating the phrase, "darkness was above the deep" (Gen. 1:2), he explains that air—which he has already equated with darkness (29)—"has covered and completely filled all the yawning and desolate and empty *khora* that reaches down to us from the regions around the moon" (32). His use of the word *khora* suggests an identification of the oceanic depths of Genesis with the *khora* of the *Timaeus*—all the more intriguing because it is not reducible to substance or matter in the context of a noetic creation. We recall that Timaeus proposes that *khora* is not altogether alien to the noetic realm: "We shall not lie if we call it a form that is invisible and unshaped [ἀνόρατον εἶδός τι καὶ ἄμορφον], all receiving, partaking of the intelligible [τοῦ νοητοῦ] in some most perplexing and baffling way" (Timaeus 51a–b). Yet Philo also insists that *khora* is empty, referring to the deep as a void, and he ignores the neighboring reference to the earth as not only "invisible" but also "unformed"—language arguably even more distinctly khoric in its resonances than "abyss." Having introduced *khora* by name, he must banish its specter, it seems. Verse 4 comes to his aid: "And the god made a separation between the light and between the darkness." A boundary is established, order is imposed, and the darkness that once filled *khora* yields to light, Philo asserts (On the Creation 33–35).

Yet *khora* seems to return in Philo's reading of the genesis of the material cosmos. In particular, it returns in the guise of the earth, which Philo associates with verses 9 to 13—"the third day" of creation. Initially all swamp, manifesting an "indistinguishable and shapeless nature," earth is subsequently separated into salt water and dry land irrigated by sweet water; once again, a boundary must be established to banish and contain the oceanic depths. "Like a mother," the earth provides "its offspring with both food and drink," notes Philo. So it is that the god "filled it with veins, like breasts, which, being opened, would pour forth springs and rivers" (39). Later in the treatise, Philo observes that human mothers with "gushing breasts" imitate the moist and nurturing Mother Earth (133). Here he goes on to describe at some length the well-watered land that is "fertile and deep-soiled" (39) and bounteous in its fruits (40–41). The latter are produced in response to divine command: "And the earth, as though it had for a long time been pregnant and travailing, produced every sort of cultivated plant, and every sort of tree, and also of fruit in unspeakable abundance" (43). Subsequently, when once again commanded to produce, earth sends forth the wide variety of land-dwelling animals (64), just as the waters bring forth water-dwelling creatures and birds fill the air (63).

"The myth of the faceless mother provides the very motivation for our exploitation of Earth, seen as 'inexhaustible matter for things,'" writes Morton, quoting Emmanuel Levinas.[51] Might Philo's earth be read as just such a faceless mother? By the time he has finished celebrating "her" marvelous bounty, Philo may risk forgetting what he has intimated earlier. The realm of becoming is marked by fragility as well as generativity: it will "sink under" the weight of too much demand. He does acknowledge that gluttony, sexual appetite, and desire for fame, wealth, and power, among other cravings, have brought a just punishment upon humans: "and the punishment is the difficulty of procuring life's necessities." As a result of human greed, the earth no longer yields its fruits effortlessly. Humans must endure the hard labor of ploughing, watering, and sowing the fields, always facing the possibility that crops will be destroyed: "Either the onslaughts of repeated rainstorms drag them down, or the weight of hail that has fallen on them snaps them off in masses, or they are half-frozen by snow, or the violence of the winds tears them up by the roots" (80). Yet Philo has no doubt that if humans should live virtuously, the god would restore the spontaneous fruitfulness of the earth (81). After all, earth's botanical bounty came into being so that humans should lack no resources for "living and living well" (77), just as its animals were intended to serve humans as their masters (83–87).[52]

But what if there is no going back to such an Eden, even if we wanted to? This is no rhetorical question, of course, when global warming has passed the tipping point. Whereas Plato's *Timaeus* offers us fragments of an eco-chorology that opens time to an as-yet-unimaginable future, Philo's strongly teleological and anthropocentric rereading of the text seems to reinscribe closure. His theology carries the burden of his totalizing ambition, proclaiming a god more powerful than Plato's, sole agent in the drama of creation. The god's surpassing power is matched, finally, not so much by the weakness or finitude as by the surpassing fertility of the earth. It is also mirrored in the human, "first citizen of the world," a creature who can claim the very cosmos as its home, having been deemed "worthy of the rule of all earth-dwellers" (142).

We might say, then, that Philo's is a light-filled, ever-green ecological vision, a cosmology that is afraid of the dark, a narrative that refuses to acknowledge its own gaps, a cosmo-politics that accommodates no rub of desire and difference. And yet *khora* continues to haunt his discourse as well. It haunts his account of the noetic creation, in which the god does not so much conceive a thought as dream a dream—a dream of darkness and the dizzying abyss, of fugitive phantoms of earth, water, breath, fiery light. It haunts his account of the material creation, in which the teeming fecundity of earth exceeds the dictates of divine command—the world an animal in travail, an animal filled with animals. It haunts his account of a fall at once all too human in its causes and eerily inhuman in its effects—the unpredictable ravages of rain, hail, snow, and wind. It haunts his account of the fragile resilience of the world of becoming—as generative complexities both destroy and give rise to ever-emergent forms. Does *khora* not, finally, also add its swaying, quaking dance to the celebration of the sabbath as "the festival, not of one city or one country, but of all the earth, . . . the birthday of the world" (89)?

Origen of Alexandria, *On First Principles* and *Against Celsus*

Writing some two hundred years after Philo, Origen (c. 185–c. 254) inherits his predecessor's commitment to the joint pursuit of biblically grounded theology and Platonizing philosophy.[53] For Origen, the latter is merely preparation for the former, and the authority of Moses clearly trumps that of Plato: "In the pure and pious soul of Moses, who rose above all that is created and united himself to the creator of the universe, there dwelt a divine spirit which showed the truth about God far clearer than Plato and the wise men among the Greeks and barbarians" (Against Celsus 1.19).[54] Yet Origen is also well versed in philosophical discourse: like his younger contemporary Plotinus, he is said to have studied with the renowned Platonist Ammonius Saccas.[55] As Elizabeth Digeser has recently argued, the ever-multicultural Alexandria nurtured a "particularly close and fruitful" set of intellectual exchanges in the early third century, resulting in "a period of tremendous innovation" that gave rise to new conceptualizations of philosophical monotheism, shared—albeit not without differences—by both Christians and non-Christians. Within a generation, when Porphyry wrote his critique of Origen and Eusebius penned his defense, boundaries of religious identity had hardened. But in Origen's lifetime it was still possible to inhabit "a borderlands region where philosophers and Christian exegetes intermingled."[56]

Origen's most explicit engagements with the *Timaeus* are found in his lengthy treatise *Against Celsus*, written late in his life at the request of his patron Ambrose. Though Origen confuses him with another Celsus who was an Epicurean, Celsus appears to have been a Platonist critic of Christianity writing in the late second century; his work, titled *True Logos*, survives only in the fragments cited by Origen in his refutation. Because of its particular polemical framing, *Against Celsus* offers a rare glimpse of Origen as an intellectual at home in the give and take of philosophical debate. I shall therefore begin with this late text and then backtrack to consider one of Origen's best-known works, *On First Principles*. Whereas *Against Celsus* allows us to place Origen in relation to traditions of interpretation of the *Timaeus*, *On First Principles* provides a fuller view of Origen's own speculations on the nature of the universe. There he translates *khora* into a new idiom of soteriological cosmology. Origen's thought is no less problematically anthropocentric than Philo's—if anything, it is more so—and yet it too bears traces of an eco-chorology.

In the fourth book of *Against Celsus*, Origen cites the following passage from Celsus's *True Logos*: "The god made nothing mortal. Whatever beings are immortal are works of god, and mortal beings are made by them. And the soul is god's work, but the nature of the body is different. In fact, in this respect there will be no difference between the body of a bat or a worm or a frog or a human. For they are all made of the same matter and are equally liable to corruption" (4.52). Origen notes rightly that Celsus here begins with "a paraphrase of the *Timaeus*" (4.54), where we read that the god himself "is the maker of divine beings, but he commanded his own progeny to make the race of mortal beings" (Timaeus 69c). But Origen criticizes Celsus for merely asserting and not "teaching" or arguing his case—thereby reversing a charge Celsus has brought against the Christians (cf. Against Celsus 1.9). Celsus has failed to demonstrate that the bodies of animals and the wide variety of plants are not in fact the "works of the god" or the creatures of "a perfect mind," as both Christians and Stoics believe. Origen's own position is that "one god is the maker of all, creating each thing for a certain purpose and reason" (4.54). He backs this up by appeal to Genesis 1. Written by "the divine spirit in Moses," its mysteries have apparently evaded Celsus. There the phrase "the god made" is applied to some parts of creation—including both humans and animals—while this phrase is lacking with respect to the genesis of other things, such as light, ocean, and plants.[57] Celsus should have considered such textual details. He should also have asked himself to whom the god's commands regarding creation were addressed (4.55). What Origen here merely hints, he makes more explicit elsewhere in this work: it is the divine *logos* whom "the father commanded in the Mosaic account of creation. . . . And when *logos* was commanded, it made everything that the father enjoined it" (2.9).[58] The father god is the creator of all things, *logos* his instrument.

At this point Origen turns his attention to the second part of Celsus's statement: all bodies are equally corruptible and thus "there will be no difference between the body of a bat, or a worm, or a frog, or a human" (4.56). An egalitarian or "flat" ontology indeed! Yet, as Origen's response highlights, Celsus's aim is not to affirm but to denigrate all bodies equally: *no* body, in his view, can be the direct creation of the god. He and Origen agree in identifying the demiurgical god of the *Timaeus* with the highest god or good. They even agree, partly, in assigning a mediating role for a subordinate, divinely begotten, deity or deities. However, for Celsus, as for Plato, the mediating deities function explicitly to distance the god from the indignity of creating corruptible bodies, whereas for Origen, as for Philo, the divine *logos* is the direct expression of the god's creative will.

In response, Origen broadens Celsus's claim so as to disrupt the particular hierarchy of immortal and mortal beings that it presupposes: as bodies, bat, worm, and frog are on the same level as the human, but so too are sun, moon, stars, heaven, "and anything else called a 'visible god' by the Greeks," Origen insists. "For the same matter [ὕλη] that underlies all things is, strictly speaking, without quality or shape" (4.56). Celsus's distinction between immortal and mortal creatures does not hold, in other words; at the same time, he is ignoring the intricate differences of creatures. All created beings have bodies of some sort, and it is the god who endows matter with the qualities that account for their diversity. Even one star differs from another, notes Origen, citing 1 Corinthians 15:39–41 in support of the enormous variety of bodies.

Bodies are not only diverse; they are also changeable. "The underlying matter is capable of receiving the qualities which the maker wills. And, god willing, a certain quality is upon this particular matter now, but subsequently, there will be another such, let's say, better and different." Here Origen refers to the transformation of the resurrection, an extreme instance of the mutability of bodies. For Celsus this mutability is a sign that bodies are not created by the god, but Origen embraces the hylic capacity for transformation. Unwavering in aligning creaturely mutability with divine intentionality, he is also unperturbed by Celsus's scoffing invocation of popular opinion to the effect that "a snake is formed out of a dead man, originating from the marrow of the spine, and a bee from an ox, and a wasp from a horse, and a beetle from an ass, and in general worms from most animals" (4.57). In a later passage, Origen refutes Celsus's charge that Christian belief in bodily resurrection is "the hope of worms" (5.14, 5.19). Enrica Ruaro suggests that Celsus here intends a similar insult to the Christian doctrine: as she paraphrases Celsus's taunt, "the worm that comes out of a dead animal is like the heavenly resurrected body that comes out of the dead earthly one."[59] In this instance, Origen appears to affirm the analogy, which suits his emphasis on the radical character of the bodily transformation that takes place in the resurrection.

By emphasizing the commonality of bodily mutability and difference, Origen proposes a kind of democracy of creatures (to adapt Levi Bryant's phrase);[60] he also "democratizes the knowledge of God," as Cinzia Arruzza puts it.[61] A relevant passage from book 7 of *Against Celsus* again opens with an explicitly marked invocation of the *Timaeus*, this time a direct citation: "Now to discover the maker and father of this universe is hard work, and it is impossible for the one who has discovered him to speak to all" (Timaeus 28c). According to Origen, Celsus takes this statement to mean that only a small number of seekers are capable of walking the path of truth, pursuing "some conception of the nameless first" by synthesis, analysis, or analogy; Celsus himself wants "to teach about what is otherwise inexpressible," but he cannot make himself understood by those who are mired in the flesh—that is, Christians. In response, Origen admits that the much-cited Platonic passage is "noble and not easily dismissed" (Against Celsus 7.42). However, he finds room to disagree not only with Celsus's interpretation but also with Plato's—or, rather, the Platonic Timaeus's—own statement. He first asserts that it is not merely difficult but impossible for any human to discover the god on his or her own. Elsewhere Origen cites Romans 11:33 to underline this point: "Oh, the depths of the riches of the wisdom and knowledge of god! How unsearchable are god's judgments, and how untraceable his ways!" As he notes in that context, the apostle "did not say that it is difficult to be able to search out 'god's judgments' but that it is not possible at all; nor did he say that it is difficult to investigate his 'ways' but that it is not possible to do so"; indeed, knowledge of god is a bottomless abyss (On First Principles 4.3.14). But Origen also insists that the god nonetheless chooses to reveal himself not to some but to all, out of *philanthropia*, or love of humanity. The divine *logos* has become flesh "so that it might be able to reach all" (Against Celsus 7.42).

In order to fill out Origen's views, let us now turn to his earlier work, *Peri archon*, or *On First Principles*, for which our primary source is Rufinus's Latin translation, *De principiis*. The title is notoriously difficult to translate, for the work deals with *archai* in several senses.[62] In the first place, there are the nine foundational principles (*elementa ac fundamenta*) of Christian doctrine, which Origen lays out in the preface of the work (On First Principles 1.praef.10).[63] Having been faithfully transmitted to all Christians from the teachings of the "holy apostles," they form the starting point of investigations for *amatores sapientae*, or philosophers. Here we see that Origen's epistemology balances inclusiveness with a degree of elitism not so far divergent from Celsus's after all. As in *Against Celsus*, he stresses that teachings necessary for salvation are made available and accessible "to all who believe, even those who seem rather dull with respect to the investigation of divine knowledge." However, he also states that further inquiry into the "ground [*ratio*]" or the "how and whence" of apostolic doctrines is left to those who are graced by the divine spirit and "prepare themselves to be worthy and capable of receiving wisdom" (1.praef.3).[64] Building on the doctrinal *archai*, such gifted teachers may hope to construct "a single body [*unum . . . corpus*]" of knowledge (1.praef.10). In addition to designating doctrinal principles, the term *archai* can also refer to the founding principles of the universe—triune god, noetic world, and sensible world, according to Origen. As in the *Timaeus*, it can refer to *beginnings* as well—to beginnings that may have to begin again: "For what was before this world or what will be after the world is not yet made clearly known to the many" (1.praef.7).

Origen does not merely intimate that the universe may itself begin again (and again): "for surely the end of this world is the beginning of the one to come" (2.1.3). He also tells the story of beginnings twice. He tells it twice not only because, like Philo, he tells of both a noetic and a material coming-to-be, but also because he gives two different accounts of the cause of this doubled genesis. One account begins with the father god and the god's *sophia*, or wisdom. The father begets *sophia* eternally and thus it (she?) is "without any beginning," temporally speaking. Although *sophia* does not *have* a beginning itself, it *is* the beginning of all things. Here Origen, like others before him (including the author of the gospel of John), reads Genesis 1:1 in light of Proverbs 8:22: *sophia* "says, through Solomon, that it was 'created [ἔκτισέ] as the beginning [ἀρχὴν] of god's ways.'" Thus when Moses writes that the god created heaven and earth "in the beginning [ἀρχῇ]," he indicates that god created "in *sophia*."[65] Elsewhere Origen brings out the erotic undertones of the Proverbs verse: the god takes joy in *sophia*, "delighted by her variegated spiritual beauty," a beauty that "summons heavenly *eros*" in all who contemplate it (Commentary on John 1.9.55). Indeed, *sophia*—who is also called *logos*, among other names—is always already pregnant with "every power and representation of the future creation, . . . formed and arranged in advance by the power of foreknowledge," "sketches and prefigurings," "beginnings [*initia*] and grounds [*rationes*] and species of the whole creation" (On First Principles 1.2.2). Origen repeats the point: *sophia* "forms in advance and contains in itself the species and grounds of the entire creation" (1.2.3); in *sophia,* "creation is always present, sketched and formed, . . . a prefiguring of what was to be" (1.4.4); "all the kinds and species have always existed, and perhaps even individual beings" (1.4.5). We note the similarity to Philo's figure of divine *logos* or reason as the seat of the noetic creation on which the sensible world is modeled (On the Creation 6.24). Yet unlike Philo, Origen explicitly rejects the concept of separately existing forms or "*ideas*, as the Greeks call them"

(On First Principles 2.3.6); moreover, his divinely prefigured creation seems already to accommodate the differentiation that he associates specifically with materiality as well as with divine delight.

Does this many-named *sophia* not, then, seem as much khoric as strictly noetic, even in its tilt toward the feminine? Drawing on the vocabulary of Proverbs 8:22, Origen later refers to *sophia* as *ktisma* (4.4.1), a term usually translated as *creature*. However, as Patricia Cox Miller points out, *ktisma* can also mean *foundation* or *building*; she proposes that for Origen *sophia* as *ktisma* is "God's binding structure in which all things are made new."[66] Indeed, one might almost designate it (her?) "mother and receptacle of the visible and in all ways sensible generated world" (Timaeus 51a). At the same time, this matrix of multiplicity and difference is also the only-begotten son and perfect image of the father and maker of all things, according to Origen. As in the *Timaeus*, *khora* queers mimesis while also enabling it. In and as the doubling of the god, Origen's khoric *sophia* bridges the gap between the god's eternal generativity and the temporal unfolding of creation's multiplicity.

Origen's second account (not unlike Timaeus's) *begins again* not with the god but with generated beings or, to be more specific, with the *fall* of generated beings into embodied existence. Origen has already broached the topic of falling before he opens this second account. In his initial discussion of the Holy Spirit, he explains that "those who are not holy by essence are made holy by participation in" the gift of the spirit. However, the very possibility that creatures might "seize bliss" seems to suggest that they might also fail to hold onto it: carelessly taking the gift for granted, they may fall from grace. That is not to say that falling is irreversible. A creature who lapses in its desire for blessedness need not "be hurled to the very depths but may return and go back to its own place and once again be able to reinstate what had slipped away through negligence" (On First Principles 1.3.8). Having entertained such possibilities, Origen acknowledges that he has digressed from his topic and returns to his discussion of the triune god, deferring further exploration of "rational nature" to its "proper place" (1.4.2). But the soteriological framing of his cosmology is already clear.

When Origen begins his "proper" discussion of rational beings, he initially stresses their variety: some are holy, some wicked, some in between—angels, demons, humans (1.5.1). Did the god create them in these different states? Definitely not: the position of every created being comes about "from merit, and not through a privilege of creation" (1.5.3). That is to say, all are created equal yet they do not remain so. Through satiation or carelessness, they withdraw from their creator, but they do so to different degrees. Fiery intellects, cooled in their ardor for the god, become souls, and souls fall into the heaviness of bodies of different sorts, placed in correspondingly diverse conditions and regions. The physical universe comes to be in and as this variegated falling. "From one beginning they are drawn in various directions according to their own individual movements, distributed through the different ranks according to merit" (1.6.2). Moreover, the process is ongoing. Origen deems it likely that "every single rational nature can, in passing over from one rank to another, come through each rank to all and from all to each, while it suffers the various movements of advancing or falling away according to its own impulses and efforts and the power of free will" (1.6.3). His is emphatically both a descensional and an ascensional bestiary, to borrow (and extend) Sallis's phrase. "Angels may become humans or demons, and demons may become humans or angels" (1.7.5). If a passage preserved in Greek can be trusted, Origen (like Timaeus) speculates that humans may also become irrational animals, even sinking as low as fish, a process that may be reversed as well (1.8.4).[67]

This raises the question of how far a rational being can descend and, alternately, how far (if at all) a nonrational being can ascend. The diversity of the world embraces not only angels, humans, and demons but also other animals as well as places and plants, Origen notes (2.1.1). The case of the other animals remains particularly unclear. Later, in *Against Celsus*, Origen vehemently denies the attribution of rationality to any nonhuman animals; like Philo in his treatise *On Animals*, Origen invokes traditional Stoic polemics to refute the Platonic argument that the universe was made as much for nonhuman animals as for humans (Against Celsus 4.74–99).[68] However, in *On First Principles*, where he is not being pressed by Celsus's critique of Christian anthropocentrism, Origen's views are less dogmatically expressed. He affirms that even fish have souls; that is, they are capable of "imagination and movement." "Bees, wasps, ants, oysters, snails"—all have souls (On First Principles 2.8.1).[69] And what is a *nous*, or rational being, but a soul that has been saved and sanctified (2.8.2)? Might an animal soul not, then, become rational? Strikingly, if also ambiguously, Origen affirms that "the majority of the nature distributed to all [rational beings] is present in animals to a certain extent, some more, some less"; hunting dogs and war horses in particular seem to him "nearly rational" (3.1.3).

While seeming to delight in the variety and mutability of living beings, Origen works hard to eliminate necessity, chance, or randomness from his account of the creation and transformation of the universe (1.praef.5; 1.8.1); we may recall how Plato's Timaeus initially attempts to eliminate necessity or the "errant cause" from his own account. Creaturely changes and consequent differences are not the result of blind fate or divine whim but are the consequence of the creatures' exercise of free will, insists Origen. Yet free will—the very factor that serves to protect the goodness and rationality of the creator god—also introduces something very like necessity, chance, or randomness into the cosmos.[70] Something very like *khora*. Why do creatures turn away from god at all, "whether through sloth or negligence" (1.5.5)? Why do some fall farther, some rise again more quickly than others? Origen has no answer to these questions, nor could he: any answer would compromise freedom, as he has framed it. As a result, his two accounts of beginnings cannot be easily harmonized. Is the diversity of the cosmos the direct effect of an omnipotent divine will, as is seemingly expressed in the figure of the *logos* or *sophia* that always already contains the "beginnings and grounds and species of the whole creation" (1.2.2)? Or is it the unintended effect of a myriad of creaturely movements, desires, and choices unfolding and coalescing in time?

To be sure, Origen attempts to hold strong affirmations of both divine *pronoia*, or providence, and creaturely free will together. He notes that the god "recalls these very creatures who differ greatly from one another in the variety of their minds to a single agreement of work and effort so that the diversity of the motions of their minds notwithstanding, they nonetheless complete the fullness and perfection of a single world and the very variety of intellects inclines toward the single goal of perfection." He explains further that the "ineffable plan" of the divine *logos* and *sophia* simultaneously preserves the free choice of all creatures and adapts the "diverse motions of their intentionalities" to the "harmony of a single world" (2.1.2). But how is this possible? In part through the wonders of divine pedagogy. Origen depicts the cosmos as a stern classroom in which some beings—demons, for example—are hostile and destructive yet may still enable those whom they beset to be made strong in adversity; at the same time, other beings—stars, for example—are avowedly altruistic, suffering sensible embodiment not as the consequence of their own faults but out of love for others. Once again making a clear reference to the *Timaeus*, he describes the universe in all its complexity as "a vast and monstrous [*inmensum atque inmane*] animal that is held together by the power and reason of the god as if by one soul" (2.1.3). Emphasis on the vastness or monstrosity of the cosmic animal is not found in the *Timaeus* (cf. Timaeus 30b). Origen's language is thus all the more striking, not only suggesting immensity of size but also hinting at the fearsome and wild.[71] The universe is not just an animal; it is a beast. As a collectivity of warring impulses and movements, the cosmos nonetheless exhibits a single, if ever-evolving, intentionality, Origen insists.

By drawing them into such close coordination despite their tensions, Origen has significantly revised theories of both providence and free will. To take the measure of this revision, it may be helpful to recall Timaeus's account of the movements of the elemental potencies that are contained in *khora*, in the guise of "nurse": "she is herself shaken by these and, in turn, being moved, shakes them" (Timaeus 52d–e). As we noted before, creative agency and affectivity are here dispersed, yet creaturely interconnectivity also produces a cumulating effect, for when the elemental potencies move *khora*, *khora*'s own motion moves them too. Moreover, the collective agency emergent in *khora* develops focus and force, and the shaking has the effect of winnowing and sorting the various kinds. Thus, in *khora*, apparently random movements produce the spacing of difference. Similarly, the movements of Origen's rational beings affect the divine *logos-sophia:* shaken by them, she shakes them in turn, we might say, giving rise to "all the kinds and species" (On First Principles 1.4.5). Of course, the resort to linear chronology is as misleading with respect to Origen's thought as it is with respect to Plato's: the differentiation of creation is always already prefigured in the eternal *logos-sophia*, in anticipation, so to speak, of the temporal movements of creatures. But clearly neither divine nor creaturely wills are absolutely free, in Origen's thought. Perhaps better: in order to create at all, the god must partly sacrifice its own freedom of will so as to grant free will to creatures.

The *Timaeus* continues to hover implicitly over Origen's discussion as he shifts his focus to the question of bodily nature as such. Origen's attitude toward corporeality is famously complex and ambivalent. It is also famously difficult to discern through the veil of Rufinus's translation, which appears to bend Origen's most controversial thoughts in a more "orthodox" direction, while Greek fragments surviving through citation by Origen's detractors may render them more "heretical." Appropriating a Pauline image,[72] Origen declares that the cosmos is in travail, laboring to deliver itself from the bondage of bodily nature (On First Principles 1.7.5). It is not altogether clear whether delivery constitutes a complete liberation from any form of corporeality or, alternately, indicates a radical transformation of the body in the resurrection; just so, it is not altogether clear whether Origen imagines rational natures to have preexisted their embodiment. If we take seriously the khoric dimension of *sophia* as the matrix of a creation that is (at least *in potentia*) always already differentiated, we may be inclined to trust Rufinus's translation: "While they were originally created as rational natures, material substance can only be separated from them by conjecture and thought, and though it seems to have been made for them or after them, they never have or do live without it; for only the life of the trinity can be rightly supposed to exist incorporeally" (2.2.2). Bodily nature, granted in advance, as it were, is what allows for not only the differentiation but also the transformation of creation (2.1.4)—traits that Origen continues to celebrate. "Corporeal nature accepts diverse and various alterations, so that it may be transformed from all things to all things." Invoking Timaeus's representation of elemental flux (Timaeus 51a), Origen seems to identify this fluidity with *hyle*, or matter—"what underlies bodies," being in itself "without qualities" yet "never found existing without qualities" (On First Principles 2.1.4; cf. 4.4.6). Later he goes on to describe the apophatic process by which the mind constricts its focus, contemplating the *punctum*, or point, of underlying matter by mentally "removing every quality" (4.4.7).

Origen departs from the philosophical tradition represented by the opinions of "great and distinguished men," as he himself notes, when he also insists that *hyle* is not eternal but divinely created, perfectly suited to "receive into itself the qualities that the god itself wanted to impose" (2.1.4). Behind this position lies a particular reading of Genesis 1:1–2a: "In the beginning God created heaven and earth, and the earth was invisible and shapeless (*invisibilis et incomposita*)." Since Genesis 1:6–10 refers to a different heaven and earth, Origen like Philo concludes that the first verse must refer to something other than the visible universe; he speculates that it might refer to an "other," spiritual world—a supercelestial "world of saints" that has its own heaven and earth—but he goes on to note that he has dealt with these verses in more detail elsewhere, presumably his lost *Commentary on Genesis* (2.3.6). He subsequently observes that many interpret the figure of the earth in these verses as a reference to *hyle*, or "unformed matter" (*informis materia*) (4.4.6). A surviving passage from the lost commentary parallels this second reading. "What kind of heaven did God create before the rest, and what kind of earth? Philo thinks that they are immaterial and intelligible creatures, ideas and models of both this earth and the firmament. Others take it that the prophet . . . indicated . . . by 'heaven' the immaterial nature, by 'earth' that which is the substance of bodies and which the Greeks call *hyle*. This interpretation is supported by the text that follows immediately: 'the earth, however, was invisible and shapeless.'" At this point, Origen makes the most of the various Greek translations of the biblical verse. "Invisible and shapeless" (Septuagint: ἀόρατος καὶ ἀκατασκεύαστος) designates matter's lack of color and quality, he suggests; "empty and nothing" (Aquila: κένωμα καὶ οὐθέν) indicates that, all receiving, it "can never be filled up" and is "devoid of anything"; "idle and disorganized" (Symmachus: ἀργὸν καὶ ἀδιάκριτον) indicates its passivity and need for order; and, finally, "speechless with amazement" (a translation apparently derived from Hebrew exegetical traditions) suggests its likeness to a soul struck with awe in the face of its creator.[73] The khoric resonances of his interpretation of Aquila's translation are especially notable.

If the god is said to impose qualities on all-receiving *hyle*, it does so in carefully calibrated response to the diverse movements of rational creatures, as we have seen. The intentionalities of creatures themselves thus partly determine their qualities as embodied souls. In its radical receptivity, *hyle* is, then, the site of creaturely affectivity, implicated in ongoing cosmic processes of contestation and cooperation, differentiation and transformation, descent and ascent. Sometimes Origen seems to imagine these processes as (very nearly) endless, as one world gives rise to another, each one different from all others. "What their number and quantity might be I confess I do not know," he writes. "If anyone can show me, I would willingly learn" (On First Principles 2.3.4). The delivery of all creation from its bondage will require "the passing of infinite and immeasurable ages," he is sure (3.6.6). The multiplicity of worlds is not only sequential but also simultaneous, moreover. Origen finds indication of this in a number of biblical passages, not least, as we have just seen, in Genesis 1:1, which seems to refer to an "other" heaven and earth. "But it is uncertain whether that world . . . is one separated from this one and far divided in place, quality, and glory, or whether it excels in glory and quality but is contained within the boundary of this world, which seems more likely to me." He is inclined to think that there is "one complete world" that contains all other worlds; "all that are and exist" remain somehow interconnected, then (2.3.6).

Origen's ecological vision lacks the vivid greens and gushing springs of Philo's. Even more than Philo, he often seems to restrict his interest in nonhuman creation either to its role in the education of human souls or to its function as an allegory for that process.[74] Where Philo's earth is in travail with the fruits of its own fertility, Origen's cosmos labors to deliver itself from the drag of materiality itself. Yet to charge Origen with "degrading nature," as does ecotheologian Paul Santmire, is surely too simple.[75] Origen's concept of humanity is almost Darwinian (*avant la lettre*, of course) in its assertion of radical mutability, and his soteriology produces a remarkably open and dynamic cosmology. His humans exist on a shifting continuum shared with angels, stars, horses, dogs, birds, fish, snails, and demons. Such human and other-than-human creatures are ever themselves, and they are also ever becoming-other. They are equal and the same, and they also shimmer with variety and diversity. They are fiery spirits, and they are also earthy, fluid bodies. They are free, and they are also entangled in the fate of the cosmos as a whole. They are learning, and they also all too often fail to learn.

Moreover, these human and other-than-human creatures are not contained by the horizon of a single, enclosed world: every ending is a beginning, and the universe—a monstrous animal—is vast and patient, beyond any creature's imagining. Origen is confident that the god ultimately promises salvation both for creation as a whole and for each individual, yet he acknowledges that even a perfected creature may fall again at any point. If the "heaven" of Genesis 1:1 yields, on his reading, an egalitarian collectivity of living beings burning with love, its "earth"—without qualities, empty, receptive, indeed very nearly nothing—hints at the ghostly mystery of their embodied existence. Both spirit and materiality are created together "in the beginning," that is, in *logos-sophia*, which provides the spacing for difference and change. Thus, the cosmos unfolds in a divine process of ongoing—indeed, perhaps infinite—adaptation to the unpredictable (if not unforeseen) actions of embodied spirits.

On the one hand, there is the agency of all creatures, albeit not the same agency for each—star, human, snail; *this* star, *this* human, *this* snail. On the other hand, there is the agency of divine *pronoia*, or providence—a coalescing of intention or purpose that is, of necessity, multiple and ever emergent, yet also (Origen assures us) unified and eternal. There is also a third: a khoric third. And the third is itself double. *Hyle* enables the embodiment of creaturely desire and difference; *logos-sophia* enables the embodiment of divine desire and unity. As the incarnation of *logos-sophia* in Christ reveals, the double is also one. And since every revelation is at the same time a withdrawal or a reveiling, Origen's dark cosmology is also a dark christology.

Athanasius of Alexandria: *Against the Gentiles* and *On the Incarnation of the Word*

In contrast to Philo and Origen, Athanasius (c. 299–273) is not known for his philosophical turn of mind. As the bishop of Alexandria in a moment marked by the advent of imperial patronage of Christianity and the accompanying turmoil of ecclesiastical controversy, he made his reputation as a staunch upholder of Nicene orthodoxy. However, Athanasius's intellectual abilities should not be too quickly dismissed. Origen's biblical Platonism formed the core of the bishop's theological heritage, and he was not ignorant of philosophy. Gregory of Nazianzus observes (while seeming to downplay the fact) that a "brief study of literature and philosophy" preceded Athanasius's Christian education (Oration 21.6), and Sozomen credits him with having been "well educated, versed in grammar and rhetoric" (Church History 2.17). Athanasius's own works appear to confirm both reports. Although the extent and seriousness of his philosophical and rhetorical education continue to be debated, even the more skeptical agree that "Athanasius appropriates the language and ideas of Greek philosophy without embarrassment, and he expresses his position easily in the prevailing terminology of Middle Platonism."[76] Here we shall focus on his two-part apologetic composition, *Against the Gentiles* and *On the Incarnation of the Word*. Apparently written relatively early in his career, when he was not yet fully consumed by his dispute with rival theologians, these works allow us to appreciate the cosmological issues that frame Athanasius's well-known concerns with soteriology.[77] Like Origen, Athanasius holds cosmology and soteriology closely together. He also modifies Origen's theological model significantly in some respects, as we shall see. Athanasius's distinctive use of the concept of creation from nothing will draw our particular attention. His thought may yield traces of *khora* where we would have least expected it, in a negation that gestures toward the fragility and mystery of all things while also foregrounding the wayward destructiveness of human beings.

We begin with *Against the Gentiles*. There Athanasius proposes to address the topics of correct worship and cosmology, pursuing "knowledge of piety (τῆς θεοσεβείας) and of truth about the universe (τῆς τῶν ὅλων ἀληθηείας)" (1.1).[78] The latter should lead to the former, in his view: that is, one who understands the truth about the universe will be led to worship its creator as the only true god. In this context, he introduces the Christian god in terms recognizably Platonic. Like the god of the *Timaeus*, his god is good (ἀγαθὸς) and therefore without envy, desiring that his creation be as like him as possible (cf. Timaeus 29e). As Athanasius frames it, "since he is good and supremely beautiful, he made the human race in his own image through his own *logos*, our savior Jesus Christ" (Against the Gentiles 2.7–8). His subsequent description of the state of humanity as originally created recalls Origen's thought: the human "continuously contemplates through his or her purity the image of the father, god the logos, according to which image the human was also created" (2.16–18); "by power of intellect [the human] clings to the divine and intelligible realities in heaven" (2.20–21); the human "rejoices in the contemplation" of god "and is renewed in its desire for him" (2.26–27). Humans are created to contemplate and desire their creator, in short.

Unlike Origen, however, Athanasius makes it clear that humans possess material, sensible bodies from the very beginning, and this provides him with an initial explanation of their fall from bliss: turning away from the contemplation of god, humans "sought rather what was closer to themselves—and what was closer to them was the body and its sensations" (3.4–5). The human soul, Athanasius explains further, is "by nature easily moved" (4.15), and it in turn moves "the limbs of the body" (4.29); it does so "not realizing that it had been created not simply for movement, but for movement toward the right object" (4.32–33). Although created for divine contemplation, humans are at the same time oddly ill-suited for their intended activity, it would seem: they are *naturally* close to the flesh and also *naturally* restless and mobile. This mobility instills in Athanasius none of the delight that it does in Origen, and a strong streak of ambivalence runs through his anthropology at precisely this point. Athanasius states clearly that humans are "rational by nature (λογικοὶ κατὰ φύσιν)" (13.28), other animals being "naturally without reason (ἀλόγων), all sorts of birds, quadripeds tame and wild, reptiles, all of the things that land and sea and the nature of the waters produce" (19.2–5; cf. 31.1–2). Yet he also insists that humans are naturally inclined to forsake rationality. Some of them fall so far that "they crawl on the earth in the manner of snails on the ground" (9.10–11).[79] (We recall Plato's "descensional bestiary" and also Origen's bidirectional version.)[80] Is humanity as Athanasius depicts it not already intriguingly khoric, bearing traces of a third kind that supplements creator and creation—a mysterious and unaccounted-for "nature" that subverts the binary of an active god and a passive universe formed out of nothing?

The Timaean resonances of Athanasius's account of creation are made more explicit in a later passage, which also sheds further light on his thinking with respect to creation beyond the human. "But the god of all is good and supremely beautiful by nature. Therefore, he is also benevolent (φιλάνθρωπος). For a good being would be envious of no one, so he does not envy anyone but wishes all things to exist so that he can exercise his benevolence" (Against the Gentiles 41.12–16; again, cf. Timaeus 29e). Echoing another Platonic work,

Athanasius goes on to explain exactly how the god exercises his benevolence. Perceiving the troubled state of created nature, ever in a state of flux and dissolution, the god "steers and steadies the whole world through his own *logos*" so that it might not return to a state of "not being (τὸ μὴ εἶναι)" (41.22–23, 41.27). The emphasis on the threat of nonexistence is distinctly Athanasian, as a glance at the Platonic text demonstrates. There the god intervenes "so that it [the shiplike cosmos] would not plunge into the boundless sea of dissimilarity" (Statesman 273d–e); chaos, not utter annihilation, is what endangers the universe, according to Plato.[81] Yet Athanasius very nearly closes the gap between Plato's chaotic flux and his own "non-being" in so far as he imbues the latter with something like agency, if only in its capacity to subvert being. Created nature is not in itself good, he suggests; it does not in itself bear the full mark of divine benevolence. "For the nature of created things, which have come to existence out of what was not (ἐξ οὐκ ὄντων), is fluid and weak and mortal when considered by itself" (Against the Gentiles 41.10–12).

Infused by the *logos*, however, the cosmos is not only stable and vibrant but wondrously well ordered. The *logos* "is present in all things and extends its power everywhere, illuminating all things visible and invisible, containing and enclosing them in itself; it leaves nothing deprived of its power, but gives life and protection to everything, everywhere, to each individually and to all together" (42.2–6). Athanasius compares the beauty and harmony of the universe to the music of a well-tuned lyre (42.22–30), the song of a well-conducted chorus (43.1–7), the movement of a well-coordinated body (43.7–12), the peaceful productivity of a well-governed city (43.12–27). "By the will and power of the commander and leader of the universe, the divine *logos* of the father, . . . everything lives and moves," he asserts (44.1–2, 44.8). Naming the wonders of the cosmos, he waxes as eloquent as Philo: "Fire burns, water chills, springs gush forth, rivers flow, seasons and periods pass, rains fall, clouds fill, it hails, snows, and freezes, birds fly, serpents glide, fish swim, the sea is sailed, the earth is sown and in its own seasons blossoms, plants grow, some are green, others ripen, others grow to old age and die, some perish and others are born and spring up" (44.8–15). Creation itself gives witness to its maker, and it is not in the parts—many of which are foolishly deified by the Greeks, he scoffs—but in the totality that the divine may be perceived: "If one were to take the parts of creation by themselves and examine each one separately . . . and isolate them from their mutual connection . . . , then one would find no one of them self-sufficient, but all in need of mutual service and only subsisting through mutual support" (27.30–37). Despite the weakness and instability of created nature as such, the divine *logos* binds it into a harmonious, interdependent whole.

It is already clear that Athanasius's thought is more tensively dualistic than Origen's, despite the many continuities. His divine *logos* must claim the fullest measure of power and transcendence with respect to creation, a power and transcendence leveraged against the inherent weakness and instability of created things. But why is the material universe so weak and unstable if it was always already the work of the ever-benevolent and all-powerful god? Rebecca Lyman observes that with Athanasius, "curiously, at the same time as divine will became primary in creation, it lost power."[82] Here too there is a kind of doubleness in the account of creation. On the one hand, the creator is weak to the very extent that its creation is weak (virtually nothing); on the other hand, the creation is strong to the very extent that its creator is strong (being itself). The strong-weak god creates in a double gesture, and the weak-strong creation reflects that doubleness. Put otherwise, the cosmos exists in and as both a tendency toward flux and dissolution—a state of *no-thing*-ness—and a capacity to take on divine orderings and harmonies—*things* and their relations.

This paradox of creation's doubleness is further articulated and indeed deepened in Athanasius's second installment, *On the Incarnation*. Cosmology in the broadest sense frames his concerns in *Against the Gentiles*, as he seeks to persuade his readers to abandon worship of what he insists are mere creatures and to worship instead the one god who is creator of all. In the second work, anthropology draws his primary attention as he seeks to provide a rationale for the incarnation of the divine *logos* as a human—an event even more central for Athanasius than for Origen. Narrowing his focus to the human creature and its divine-human savior, Athanasius also cleaves more closely to the biblical text than he does in *Against the Gentiles*.

In so doing, he begins again. "We must first speak about the creation of the universe and of its creator god" (On the Incarnation 1.31–32). The biblical account, properly understood, is to be distinguished from other, competing doctrines—the Epicurean claim that "all things came into being of their own accord and by chance" and "without providence" (2.3–6); the Platonic teaching that "God made the world from pre-existent and uncreated matter" (2.16–18); and the heretical assertion that "the maker of all things" is different from the father of Jesus Christ (2.33–34). In rejecting the Platonic teaching, Athanasius insists that if the god "is not the cause of matter but simply makes things from existing matter, then he is found to be weak, unable to fashion any of the things that exist without matter" (2.21–24). Athanasius's own god is not weak but strong—no mere craftsperson but a powerful creator. He repeats that "through the *logos* the god brought the universe, which previously by no means and in no way existed (μηδαμῇ μηδαμῶς ὑπάρχοντα), into being out of what was not (ἐξ οὐκ ὄντων)" (3.4–6). In support of his position he adduces three authoritative passages, the first and most important of which seems to stand in metonymically for the whole of Genesis 1–3: "In the beginning God made the heavens and the earth" (Gen. 1:1). (The other two proof texts are swiftly dispensed with—the *Shepherd of Hermas* mandate 1 and Hebrews 11:3). Having cited the first verse of Genesis, Athanasius repeats the Platonic point that "the god is good" and "being good, has no envy for anyone," concluding that "envying nothing its existence, he made all things out of what was not through his own *logos*" (On the Incarnation 3.13–16). He will go on to offer a selective paraphrase of the narrative of Genesis 1–3. However, by ending his direct citation after the first verse of the first chapter, he is able to avoid any mention of the troubling elements introduced in verse 2: "but the earth was invisible and unformed (ἀόρατος καὶ ἀκατασκεύαστος), and darkness was over the abyss, and a divine wind was being carried along over the water" (Gen. 1:2; LXX). As we have seen, Origen (among other readers) takes the phrase "invisible and unformed" to be an indicator that the "earth" of Genesis 1:1–2 refers to unformed matter, whether divinely created, as Origen suggests, or uncreated, as Hermogenes is said to have argued.[83] Thus, Athanasius secures an austerely apophatic cosmology not only through his explicit and hyperbolic negations ("by no means and in no way existing," "out of what was not"), as in *Against the Gentiles*, but also through his implicit *unsaying* of the very verse in scripture that might seem to introduce ambiguity. The doctrine of a "strong" divine creation out of nothing remains axiomatic for him, as does the corresponding (if also paradoxical) assertion of the frailty and corruptibility of created nature. The bare existence of the material world is the gift of an inherently generous god whose creative and sustaining beneficence, manifest in and as the divine *logos*, staves off an otherwise inevitable return of all things to nonbeing.

In the case of human creatures, the *logos* comes as a special gift. "And among these creatures, of all those on earth he took pity on the human race" (On the Incarnation 3.16–17) Why? It would appear that humans are somehow exceptional—exceptionally frail, that is. Indeed, the god perceives that they will not be capable of "remaining in conformity with the *logos* of their own generation"; thus he makes them different from "all the irrational animals (ἄλογα ζῶα) upon the earth" (3.17–20). This difference is explained biblically: "The god created them according to his own image" (3.20–21; cf. Gen. 1:27). Athanasius glosses the text, explaining that the god "gave them a share in the power of his own *logos*" (3.21)—something beyond the *logos* of their own generation, then. It is as if they possess ghostly doubles (σκιάς) of the divine *logos*, he suggests, rendering them *logos*-like or rational (λογικοί), unlike other animals. The god intends that this gift of rationality should render humankind capable of "remaining" in the blessed state in which they are created (3.22–24). For extra measure, the god also gives them a place to live—paradise—and a law to follow, making the consequences of disobedience—expulsion from paradise—very clear (3.25–34). Yet these gifts too prove insufficient owing to the inherent instability of the human will, which appears to be haunted by its virtual nothingness, just as the mortal flesh is; again, we note a contrast with Origen's more optimistic thought.[84] Humans "no longer remained as they had been created" (4.15–16), Athanasius observes; yet paradoxically by "not remaining" they also cleave closer to their nature. "For the transgression of the commandment turned them to what was natural, so that, as they had come into existence from not being, so also they might similarly suffer in time the corruption suited to their non-being" (4.17–20). To be corruptible, both morally and ontologically, is human nature, then. Is this the ultimate consequence of divine creation out of nothing, compounded by the all-too-subvertible gifts of rationality and law?

Of course, for Athanasius, this cannot be the last word. The good and all-loving god performs one last, heroic act in an effort to rescue humanity from utter ruin. It is an act of self-sacrifice: the divine *logos* becomes human, subjecting itself to mortal flesh so as to restore humanity to its blessed state. The purposes of this incarnation are twofold: to satisfy and thus overturn the penalties resulting from the transgression of the law and to renew and fulfill the promise of the divine image with which humanity was initially graced. "And he did this in his benevolence, . . . in order that, as humans had turned to corruption, he might turn them back again to incorruption and might give them life for death, in that he had made the body his own" (8.28–34). Humans, it seems, cannot ultimately save themselves. The very characteristics that make them distinctly human—not least rationality, language, consciousness, and a capacity for self-transformation—also render them vulnerable to self-destruction. What *can* save them? Nothing less than an all-powerful god: Athanasius famously opts for a top-down model of salvation. But there might be another way to angle his soteriological model. *What* can save humanity? Nothing less than to be caught up in the most intimate embrace possible with the divine *logos* that, as we have seen, binds all of the "parts of creation" together, "all in need of mutual service and only subsisting through mutual support" (Against the Gentiles 27.30–37). Nothing less than to be made aware that "everything is interconnected," as Timothy Morton frames what he calls "the ecological thought."[85] From this perspective, the problem for humans is not so much their mortality (as Athanasius has it) as it is their refusal to acknowledge their transience and dependence on other beings (as Athanasius's thought also implies).

Athanasius ultimately returns to some of the themes sounded in *Against the Gentiles* with regard to the vivifying presence of the divine *logos* in the cosmos as a whole. Given the anthropocentrism of his topic in *On the Incarnation*, he does so primarily by way of explaining how the *logos* can come to be present in a single human body. If the *logos* is everywhere, then it can also be anywhere, he suggests. "If, then, the *logos* of the god is in the world, which is a body, and it has passed into it all and into every part of it, what is astonishing or what is unfitting if we say that it also passed into a human?" he asks (On the Incarnation 41.19–22). "The one who admits and believes that the divine *logos* of the god is in all and that everything is illumined and moved by it, would not think it unseemly that a human body also should be moved and illumined by it" (42.5–8). True, it might have revealed itself in the body of "the sun or moon or stars or fire and air"; indeed, this would have been more seemly, Athanasius acknowledges, with a nod both to Plato and to traditional Greek piety (43.1–3). Yet nonhuman creatures did not need "a healer and teacher" (43.6); only humans did. Athanasius stresses the point: "Nothing then in creation was in error in its ideas about god except the human alone"; sun, moon, sky, stars, sea, and air, "knowing their own creator and ruler the *logos*, remained as they had been made" (43.10–14). Almost as if anticipating the late modern concept of the anthropocene in his scripting of salvation history, Athanasius marks humans as exceptional in their capacity to sin, precisely by losing sight of the divine *logos* that orders the universe as a harmonious totality.

Paradoxically, then, Athanasius's very anthropocentrism yields an ecologically pertinent insight—even if it does so at the cost of distinguishing humans sharply from the rest of "nature." (This is a problem shared by the concept of the anthropocene, as it happens.) Even more intriguing, from an ecological perspective, is his articulation of a distinctly apophatic cosmology—that is, a cosmology anchored in the negation of cosmology as such. For Athanasius, to say that some*thing* exists is to say at the same time that it does not exist, that it is no*thing* at all. Put otherwise, all *things* are in a state of flux, dissolution, corruption. This invocation of the radical instability of the universe recalls the Platonic *khora*: created nature, as Athanasius perceives it, has much in common with the "errant cause." Admittedly, an apophatic cosmology may here seem to be merely the complement of an all-too-affirmative doctrine of god. For Athanasius, however unstable and even chaotic creation may be *by nature*, it is ultimately stabilized and harmoniously ordered by the divine *logos*. Divine power subdues nature, then—including human nature. Yet we have also seen that Athanasius's famously strong god is haunted by weakness, just as his creation turns out to be less fully submissive than we might have expected. The ex nihilo, as he interprets it, not only reinstates something very like *khora* at the heart of the nature of nature, so to speak, but also potentially compromises the god's asserted omnipotence.

A demiurgical deity, given even imperfect materials to work with, might have achieved a level of relative mastery, but this god is lord of *nothing*, giver of *all*. Creation is only fully redeemed when the god has submitted itself to the frailty of flesh and then sacrificed that very flesh: if at the beginning there is creation from nothing, at the end there is the humiliating death on the cross (On the Incarnation 20–26). In between, in the midst of time, the world of becoming emerges in the tension and play between the radical mutability and corruptibility of created nature as such and the transcendent orderings that sustain both otherness and interdependence. Creation is always itself and always also more than itself, always also exceeding itself, in other words—at once fragile and strong, disordered and harmonious, fragmented and interconnected, next to nothing and very nearly divine. Of course, Athanasius strives mightily to uphold the radical difference, even opposition, between the being of god and the nonbeing of creation, an opposition that places great strain on his christology. I am suggesting, however, that the difference does not fully hold. And where it fails to hold, we discern glimmerings of an eco-chorology.

Augustine of Hippo: *On Genesis*

Turning to Augustine (354–430 CE), we leave Alexandria behind. Yet we do not fully escape the gravitational field of its theological culture. A Latin translation of Athanasius's *Life of Antony* plays a momentous role in Augustine's famous conversion, as he narrates it in his *Confessions* (8.6.15). At the same time, the theological teachings for which Athanasius was best known seem to have had relatively little impact on the younger theologian. As David Brakke notes, despite Athanasius's "weighty personal reputation" in the area of trinitarian theology in particular, his "influence on Augustine's thought was indirect and less substantial than might be supposed."[86] Origen's influence, in contrast, was significant. It was partly mediated by Ambrose and other Latin-speaking theologians whose knowledge of Greek was better than Augustine's. However, Augustine also had direct contact with Latin translations of Origen's works; the questions are what and when. György Heidl has made a strong, if necessarily speculative, case that translations of Origen—in particular his *Homilies on the Song of Songs* and his *Homilies* and *Commentaries on Genesis*—significantly influenced Augustine's thought at a very early stage. Indeed, argues Heidl, Origen's writings appear to have played a crucial role in Augustine's conversion from Manichaean to catholic Christianity.[87] Heidl suggests they would have numbered among the *libri pleni* to which Augustine refers in an early dialogue (Against the Academics 2.2.5).[88] As such, they came to supplement and partly displace the *libri platonicorum* that initially set Augustine on the path of a philosophical life (Confessions 7.9.13, 7.20.26).[89] Having read the books of Origen and other Christian writers, which he characterizes as "full" or "complete," Augustine realizes that the "books of the Platonists" are incomplete by comparison, lacking the full revelation of Christ as divine *logos*.[90]

The influence of Origen's younger Platonic contemporary Plotinus is significant and Augustine acknowledges as much. However, it is to Origen that he seems to owe the strongest debt with respect to the cosmological reflections evoked by his reading of the first chapter of Genesis. In fact, Augustine's interpretations of divine creation ex nihilo are arguably more Origenian than Athanasius's are. Khoric stirrings may be detected across a range of exegetical works, from *On Genesis Against the Manichaeans* (388–389 CE), to the *Unfinished Literal Commentary* (393–394 CE), to the *Literal Meaning of Genesis* (401–415 CE) and—in a different genre—the famous *Confessions* (397–401 CE). *Confessions* offers Augustine's most distinctive and expansive reading of the first two verses of scripture, yielding two enigmatic figures that straddle the borders of eternity and time, god and world—namely, the "heaven of heavens" and unformed matter. In this reading, as Catherine Keller puts it, "a counter-ontology seems to emerge."[91] Yet Augustine's commentaries also offer rich musings on these verses, and it is with these that we shall begin.

"*In principio fecit deus coelum et terram*: in the beginning god made heaven and earth." This is the first verse of scripture in the Latin version cited by Augustine. Like Origen, he identifies the "beginning" of Genesis 1:1 with Christ, paraphrasing John 1:1–3 in support of his reading: "since the *logos* was with the father, through which and in which all things were made" (On Genesis Against the Manichaeans 1.2.3).[92] In his first Genesis commentary, written to refute the Manichaeans, he also identifies the "heaven and earth" of verse 1 with "the whole of creation," not yet actualized but in its potentiality as "basic matter, unsorted and unformed (*primo . . . materia . . . confusa et informis*)." He adds, "I think the Greeks call it *chaos*" (1.5.9). Whatever one calls it, it is "made from absolutely nothing (*de omnino nihilo*)": like Athanasius, Augustine insists that the god's omnipotence would be compromised if the god were dependent on preexisting matter. He explains that scripture refers to this divinely created proto-cosmos as heaven and earth "not because that is what is already was, but because it was able to be that"; put otherwise, it is "a kind of seed of heaven and earth" (1.7.11).

Like Plato's *khora*, Augustine's *primo materia*, his *chaos*, is known by many names. In addition to the "heaven and earth" of verse 1, the second verse of scripture refers to it as "the earth invisible and shapeless and the abyss with its darkness," as well as "the water over which the spirit was being borne." A multiplicity of names is needed lest the reader take any of them literally: "If only one were used people would think it was just that one thing they normally understood by that term." This third thing—not god but not quite cosmos either—eludes language, then. Yet language can gesture toward it. "Heaven and earth" indicates its character as sheer potentiality: "Heaven and earth were going to come from it." "Earth" and "abyss" indicate that it is not perceptible by the senses. "Water" suggests its malleability and receptivity with respect to the shaping of the divine demiurge. It also hints at nurturing capacities: since "everything that is borne on the earth, whether animals or trees or grasses and anything else of that sort, starts off by being formed and nourished from moisture" (1.7.12).

In the *Unfinished Literal Commentary on Genesis*, written just a few years later, Augustine reiterates his threefold exegesis: "We name one thing in three ways, as the material of the world, as formless material, and as workable material" (4.15). Yet he also raises new questions: is the thrice-named *materia* the *materia* for the visible or corporeal creation alone or also the *materia* for the invisible or intellectual creation? This issue is not addressed in the earlier commentary.[93] Here he speculates that if *materia* points toward the visible or corporeal creation alone, then the "spirit of god" referred to in verse 2 as "being borne over the water" may indicate "a created vitality by which this whole visible world and all bodily things are held together and activated, a vital force to which the almighty god has granted a kind of power to serve him by acting in the things that are being produced." Something like Timaeus's "world soul," then. If, however, *materia* points toward both invisible and visible creations (as Origen had also suggested, with reference to the "heaven and earth" of verse 1), then the "spirit of god" would necessarily refer to the "unchanging and holy spirit" of the divine trinity, Augustine opines (4.17). That is, the passage would signify the demiurgic activity of the god itself, bestowing on the material "its appearance and forms" (4.15). But a third interpretation also suggests itself to Augustine, and this too has distinctly Timaean resonance. He proposes that the first two verses of Genesis may, alternately, refer to proto-elements of heaven (fire), earth, water, and spirit (air)—"not that they were already distinguished from each other, but that in the as yet shapeless mishmash (*informi confusione*) of that basic material they were still being marked out beforehand as due to originate from it" (4.18).[94] Augustine leaves all three of these possibilities in play.

Delight in the multiplication of interpretive possibilities is even more evident in his third and most ambitious Genesis commentary, *On the Literal Meaning of Genesis*. "We read in the divine books such a multitude of true meanings," he writes. Confronted with this plenitude of possibilities, his questions pile up; indeed, he acknowledges that a reader might ask, "Why is practically everything hidden still in a heap of questions?" (Literal Meaning 1.21.41). Nonetheless, a coherent theory of divine creation does emerge from Augustine's inquisitive and open-ended process of interpretation. Crucial is his distinction between three aspects of creation: as eternally present in the divine *verbum* or *sapientia*; as simultaneously and instantaneously created with time itself; and as unfolded within time. The scheme is recognizably Origenian, and yet Augustine has added his own touches. In the first instance, creation exists in word or wisdom as a kind of divine foreknowledge. Augustine refers to the "unchangeable plans/thoughts/causes (*rationes*) for all creatures in the word of god" (5.12.28; cf. 1.18.36) and references John 1:1–3: "In the beginning was the word and the word was with god and god is what the word was. This was in the beginning with god. All things were made through it and without it was made nothing" (5.13.29). In the second instance, creatures exist "both in god and in themselves" (5.18.36), as seeds planted "in the roots of time, so to speak" (5.4.11). This is the aspect of creation that Augustine understands to be narrated in Genesis 1 and further affirmed in Sirach 18:1: "he created all things simultaneously" (5.17.35). In the third instance, god "unwinds the ages which he had as it were folded into the universe when it was first set up" (5.20.41). The planted seeds now grow. The beginnings of this temporal unfurling, in which divine providence remains active, are described in Genesis 2, according to Augustine's reading. "And so creatures, once made, began to run with their movements along the tracks of time" (5.5.12).

The first two verses of Genesis 1 continue to exert a particular fascination over Augustine. He is convinced that "in the order of causes, not of time, . . . the first thing to be made was formless and formable matter, both spiritual and corporeal"; this matter is, of course, divinely created out of nothing (5.5.13). Here as in his previous two commentaries, he cites Wisdom 11:18 to buttress the argument: "You who made the world from formless matter" (1.14.28, 5.17.35).[95] However, the Genesis text remains primary. Again, Augustine considers a range of possible meanings of "heaven and earth," "invisible and shapeless," "darkness," and "abyss." Speaking in multiple voices of formless matter, the scripture nonetheless conveys one truth, it would seem. Augustine clarifies: the phrase "in the beginning" suggests that the divine *son* is "the start (*exordium*) of creation still in its formless imperfection" while god's speech in verse 3—"Let light be made"—suggests that the divine *word* "confers perfection on creation by calling it back to himself." Divine creation in its second aspect, *qua* potentiality, thus constitutes a double gesture—as formless and as formed. In both cases, god creates ex nihilo in the son, word, or wisdom; however, the name *verbum,* or word, is only properly invoked in the second case, Augustine now explains. As formless and incomplete, creation "tends toward nothing (*tendit ad nihilum*)"; as formed and complete, it responds to the call—the *word*—of god, turning toward its creator so that it may imitate "in its own measure the form of the word which adheres eternally and unchangingly to the father." The divine word calls and informs. The formless *materia*—whether corporeal or spiritual—turns, imitates, and is formed, "every element in its own way" (1.4.9).

Why does Augustine describe creation as a kind of call and response rather than simply as a divine fiat? Is it because the response is not wholly determined by the god? What is formless "does not imitate this form of the word if it turns away from the creator and remains formless and imperfect," he notes (1.4.9). "A creature, . . . even a spiritual and intelligent or rational one, . . . can have an unformed life," he also observes (1.5.10). Some creatures do not *turn* and *imitate* the word, then. Not unlike Origen, Augustine here seems to write freedom and openness into the very fabric of creation. The formlessness inscribed in the first two verses retains a kind of agency in its capacity not only to con-form but also to de-form, with respect to the divine *verbum*, form itself. And this is evidently true not only of temporalized creation, or creation as humans know it, but also of creation as only angels know it: "primordially, so to speak, or originally, just as god first fashioned it" (5.4.10). As god first fashioned it: all at once, in and as the beginning of time. In and as the utterance of the divine word, then? Yes and no. Divesting itself of speech, the divine word *babbles* formlessness, we might say (extrapolating from Augustine's metaphors). It also *speaks* the formation of what is unformed and thereby perfects or completes creation—albeit necessarily imperfectly, incompletely. In the beginning, then, all was already said and done, and at the same time already partly unsaid, undone.

Creation is a babbling-speaking. It is heaven and earth—spiritual and corporeal, formed and formless. It is invisible, dark, deep, brooding (1.18.36). It is light and dark, sky and earth, land and water, seeds and plants; it is living souls that crawl and fly and swim and walk. As he works his way through the verses of Genesis 1, Augustine interweaves his own observations of other creatures: "It is . . . absolutely certain that fishes have memory. This is something I have myself experienced, and anyone who wants to can experience it too." As for birds, "we can observe their way of life with our own eyes to be both governed by memory and very talkative and to display great skill in the building of nests and the rearing of their chicks" (3.8.12). And what of the tiniest animals, many of which, Augustine opines, emerge from rotting things? "All things . . . have in them a certain worth or grace of nature, each of its own kind." Divine wisdom grants "even the very last and least of things . . . due shape and form." Some of them may fill humans with horror because we fear our own mortality. However, "if we pay close attention we are more amazed at the agile flight of a fly than at the stamina of a sturdy mule on the march; and the cooperative labors of tiny ants strike us as far more wonderful than the colossal loads that can be carried by camels" (3.14.22). There is both inevitable vulnerability (witness the food chain!) and a marvelous persistence in all things: "We see them, from the biggest elephants down to the smallest little worms, doing whatever they are capable of . . . to safeguard their bodily, time-bound health and welfare" (3.16.25). Humans are distinct in having been given both the gift of reason (3.20.30) and the capacity for sin (3.24.37). Underlining the latter, Augustine notes that of humans alone it is not said that they are good in themselves but only as part of a "very good" totality. He surmises that scripture wants to emphasize that humans are only good "when rightly coordinated with the whole, with the universe (*recte ordinata cum toto atque uniuerso*)." He is, however, confident that divine providence sustains the beauty of the universe itself, even when humans fall out of coordination, as they inevitably do: "Even if things individually become deformed by transgressing, nonetheless the totality (*universitas*) together with them in it remains beautiful" (3.24.37).

Creation is a babbling-speaking. It is also a nothing-something, as Augustine names it in his famous *Confessions*. "How, god, did you make heaven and earth?" he initially asks. There was as yet no "where" in which to make them, no "something" from which to make them, Augustine muses. He concludes, "Therefore you spoke and they were made, and you made them in your word" (Confessions 11.5).[96] No mere demiurge or artificer, his god *speaks* "all things simultaneously and everlastingly" (11.7). Augustine marvels, "Who understands it? Who will explain it?" His heart is pierced as he contemplates the first verse of Genesis and the figure of divine *sapientia* or wisdom harbored within its "beginning": word's feminine alias. He is beset by an intensity of contradictory feelings. "*Et inhorresco et inardesco*: and I shudder and kindle" (11.9). Divine wisdom inspires both fear and love.

If the figure of *sapientia* fills Augustine with ambivalent emotions, the figures of "heaven and earth" also found in that first verse likewise evoke strong affective responses. Where heaven inspires his unbounded admiration, earth fills him with something like horror, as we shall see. He continues to sustain a multiplicity of possible interpretations but here, in the context of the *Confessions*' later books, he is inclined to stress the particular relation of the primordial creation to temporality. Neither coeternal with god nor cotemporal with the cosmos, heaven and earth sit on the border of eternity and time—but for opposite reasons, so to speak. Emphasis falls on their opposing qualities. "Out of nothing you made heaven and earth," he addresses his god, "a great thing and a small thing, . . . one near to you (*prope te*), the other near to nothing (*prope nihil*)" (12.7). Changeless form and unchanging formlessness—both of these initial creations evade "the whirling vicissitude of time" (Confessions 12.9). Conflating "heaven" with the "heaven of heavens" of Psalm 15:16, Augustine identifies it as "some intellectual creature" that holds its creaturely mutability in check "through the sweetness of the most happy contemplation"; by thus cleaving to the divine, it sustains a wondrous stability of form. This curious figure has much in common with Origen's soul of Jesus, likewise uniquely exempt from the fall into bodily mutability (On First Principles 2.6.3). At the same time, it recalls Origen's suggestion that the "heaven" of Genesis 1 might refer to a supercelestial "world of saints" (2.3.6), for Augustine also refers to the "heaven of heavens" as a divine city or house and as a chamber to harbor the soaring songs and wordless groans of his love for god (Confessions 12.15–16). Heaven is not exactly a place any more than it is a person. Yet it does seem to open a space of potent possibility for Augustine, a space that may harbor its own khoric stirrings.

As elsewhere, Augustine associates the "earth" of verse 1 with both the "invisible and unformed earth" and the "deep" and "darkness" of verse 2. "Before you formed and differentiated that unformed matter, there was not anything, not color, not figure, not body, not spirit," he confesses to his god. "Not however absolute nothing (*non tamen omnino nihil*): there was a certain formlessness without any appearance/shape/beauty (*sine ulla specie*)" (12.3). As Keller notes, "It is Augustine's preoccupation with the 'certain formlessness,' his pluralist meditation upon the complexity of this tiny text, which distinguishes his exegesis."[97] Like *khora*, this formlessness, which is neither something nor nothing, evades both sense and intellect. How then can it be known at all? The closest human thought can come is to attempt "either to know it by being ignorant of it, or to be ignorant of it by knowing it (*vel nosse ignorando vel ignorare noscendo*)," suggests Augustine (12.5). Here he may have in mind Plotinus's gloss on Plato's "bastard reasoning": "What wants to be a thought (νόησις) is not a thought, but a sort of thoughtlessness (ἄνοια). . . . It thinks a dim thing dimly and a dark thing darkly and it thinks without thinking" (Ennead 2.4.10). Augustine describes his own attempt to imagine formlessness, which results in the vision of a monstrous shuffle of "countless and varied shapes (*species*)." "My mind turned up forms (*formae*) that were hideous and horrifying, appearing in confused order, but forms nonetheless; and I called it formless not because it lacked form but because it had one so bizarre and incongruous that, if it had appeared before me, my senses would have recoiled and I, in my human frailty, would have been deeply disturbed (*conturbaretur*)" (Confessions 12.6). Imperfect as Augustine's attempt is, his affect suggests that he has apprehended something of matter's indeterminacy: as Plotinus puts it, "the soul . . . is pained by the indefiniteness [of matter] (ἀλγοῦσα τῷ* ἀορίστῳ), as if by the fear of being outside the realm of beings, and it cannot bear to stay in non-being for long" (Ennead 2.4.10). When considered from the perspective of materiality, the cosmos is a little creepy, in other words.

As Augustine continues to meditate on the shifting forms tossed up by his imagination, he comes to realize that changeability itself must be the invisible formlessness that he has been trying vainly to visualize. A groundless ground indeed. If only he could capture the moment of transition from one form to the next—if only he could perceive that which is "capable of receiving all the forms into which changeable things are changed." But he cannot—not quite. "And what is it?" he asks. "A soul? a body? a figure of a soul or body?" He concludes, "If one could say 'a nothing-something' and 'an is is-not,' I would say this is it" (Confessions 12.6). He sums up his god's creative process: "You made the world of unformed matter, which you made from no thing (*de nulla re*), it being almost no thing (*paene nullam rem*)" (12.8). Thus, between the nothing of the ex nihilo and the something of the cosmos—between eternity and temporality—is the almost-no-thing. The nothing-something.

This mysterious figure seems just barely to break the surface of scripture's polysemic abyss. That fathomless deep, not unlike the nothing-something itself, evokes both Augustine's wonder—"*mira profunditas!*"—and his dread: "It is a horror to look into it, a horror of honor and a trembling of love," he cries at the end of his interpretation of the primordial "earth" (12.14). Why does he allow himself to be drawn so deeply into these two little verses? Why does he make so much of what Athanasius perhaps wisely ignored—namely, the elusive trace of a precosmic materiality? (Then again, even Athanasius's more austere "nothing" did not eradicate it altogether.) Keller suggests that the nothing-something acknowledges "the chaosmic spontaneity of all becoming"—in her opinion an acknowledgment that Augustine will subsequently suppress. She adds, "It is the indeterminacy of a freedom to actualize good or ill," signifying "the potentiality for the whole gamut of creaturely choices."[98]

Indeed, "good or ill" and "choice" may be terms too determinate to capture the subtle agency or affectivity conveyed by Augustine's poetically paradoxical rendering: after all, the nothing-something can be known only in a kind of unknowing, by stripping away all of our usual modes of sensory perception or intellectual cognition. In the beginning there is mutability; the cosmos is a flow. But becoming is not only a flow. It is also all the turnings toward form—the ongoing coalescing of beings, objects, things from within the flow. In the *Literal Meaning of Genesis*, he describes these turnings as a response to the divine word that evokes the potentiality of myriad creatures and harmonious orderings. In *Confessions*, he also associates the turnings with the figure of the "heaven of heavens," which seems to represent a kind of limit case of absolute conformity—or put otherwise, of absolute receptivity of form. In this version, then, both "earth" (the nothing-something) and "heaven" (the heaven of heavens) bear khoric traces, conveying the fluidity and responsiveness at the creative heart of materiality. "Heaven and earth" are themselves enfolded in the divine beginning of Genesis 1:1—the babbling-speaking that is the khoric matrix of all creation. Neither sheer flux nor absolute stability of essence makes a world; rather, the temporal realm of becoming emerges in the play of flux and stability.

Like Origen, Augustine—the author of the doctrine of divine predestination!—writes a surprising degree of freedom and openness into the cosmos. Whereas Origen inscribes that openness in a speculative cosmological narrative of fallings and ascents, Augustine discovers it through an exegetical obsession that unfurls into a theory of time, while also giving rise to a theory of materiality. *Beginning, heaven, earth*: these figures straddle the border of time and eternity, as Augustine reads them, a border that thickens with possibility. Indeed, that border swells from a small first verse to a full chapter in the biblical text, giving birth to a whole world. Like his Alexandrian forerunners, Augustine is not content with a strictly demiurgical model of creation. What might seem to be the vestige of a demiurgical model—pre-cosmic matter—is in fact merely borderline stuff. His nothing-something is a third kind, we might say, neither god nor world. How do we recognize it? Because it is so hard to see. And because it leaves us, like Augustine, feeling so very unsettled. If "the uncanny and uncertainty are basic to the ecological thought," as Morton has it, then Augustine is already beginning to think ecologically.[99]

The Rabbis: *Genesis Rabbah*

Unlike the other works we have discussed in this chapter, the rabbinic commentary known as *Genesis Rabbah* is not a single-authored text but a compilation. Most likely redacted in early fifth-century Galilee,[100] it is roughly contemporaneous with Augustine's commentaries, and like Augustine, the rabbis are particularly fascinated by the first two verses of the biblical text. To what extent do the rabbinic and Christian commentators participate in a common exegetical culture? Scholarly opinion is somewhat divided on this point, and the histories of dialogue, debate, and borrowing for the most part elude us. Comparing interpretations of Genesis 6:1–4 in Augustine's *City of God* and *Genesis Rabbah*, Annette Yoshiko Reed notes that some hermeneutical convergences "cannot be explained in terms of direct contact" yet also cannot be satisfyingly accounted for culturally with reference to a lowest common denominator, so to speak.[101] By the same token, late ancient exegetical texts may provide evidence of shared cosmological sensibilities extending beyond known lines of influence. For the rabbis as for their Christian contemporaries, there is mystery at the heart of creation; yet for the rabbis the mystery is specifically marked as a topic of secret or even forbidden study.

Our focus will be on the first chapter of *Genesis Rabbah*, which begins with a series of *petihot*, exegetical discourses that proceed by bringing other biblical verses to bear on the passage at hand—in this case, the first verse of Genesis. Each is woven from the cited sayings of a revered rabbi or rabbis. The initial *petihah* is opened by Rabbi Ohaya's citation of the first line of Proverbs 8:30–31: "Then I was beside him like a little child. . . ." Ohaya observes that the letters for "child" can also be construed as "teacher," "great," "covered over," "hidden," "worker" or "work-plan" (Genesis Rabbah 1.1.1).[102] With this multiplication of exegetical possibilities, a range of images of divine *hokmah*, or wisdom, are evoked. In the same passage from Proverbs, *hokmah* also says, "The lord made me as the beginning of his way" (Prov. 8:22); thus, "child" and all its other possible construals can be understood as interpretive glosses of the "beginning" of Genesis 1:1: "in the beginning god created." "Work-plan" in particular suggests an identification of the beginning with divine *torah*, and the biblical verse is revoiced as follows: "The *torah* speaks, 'I was the work-plan of the holy one, blessed be he.'" An explanatory analogy is offered: "In the accepted practice of the world, when a mortal king builds a palace, he does not build it out of his own head, but he follows a work-plan" (Genesis Rabbah 1.1.2). As Peter Schäfer points out, this remarkable interpretation "presents the Torah as the blueprint of the world in the double sense of the master plan that God follows in creating the world but also of the document that contains in it, *in nuce*, everything that will eventually unfold during the process of creation in its full historical dimension."[103]

The comparison of god to both a king and an architect is also found in Philo's *On the Creation*: like a royal architect who, in founding a city, first conceives a mental image of that city, the divine demiurge first creates a noetic paradigm on which the sensible universe is modeled (On the Creation 19). However, Philo emphasizes that the "cosmos of ideas" is a strictly mental production that "would have no other place than the divine *logos*" (20), whereas *Genesis Rabbah* represents the divine *torah* as an externally inscribed text: "The architect does not build out of his own head, but he has designs and diagrams. . . . Thus the holy one, blessed be he, consulted the *torah* when he created the world" (Genesis Rabbah 1.1.2). In this respect, the rabbinic text is closer to Philo's Platonic prototype: "According to which of the models did the architect construct the cosmos? . . . If the cosmos is beautiful and its demiurge good, it is clear that he looked toward the eternal" (Timaeus 28c–29a). Henry Fischel suggests that either the Platonic passage or its "rhetorical descendants" may have inspired the midrash.[104] More recently, Niehoff has proposed an alternative, namely, that *Genesis Rabbah* may here be influenced by Origen, who spent the last two decades of his life in Palestinian Caesarea. Origen is not only the possible mediator of the Philonic figure of the architect but also the author of a rival interpretive tradition: whereas Origen identifies the "beginning" of Genesis 1:1 with Christ the divine *logos*, the rabbis insist otherwise: "And the word for 'beginning' refers only to the *torah*, as scripture says, 'The lord made me as the beginning of his way'" (Genesis Rabbah 1.1.2).[105]

"Hidden" is one of the proposed meanings for the noun initially vocalized as "child" in the first *petihah*. The themes of hiddenness, silence, and secrecy are picked up in the next discourse, which opens with Rabbi Huna's citation of Psalm 31:19 in the name of Bar Qappara: "Let the lying lips be made dumb which arrogantly speak matters kept secret against the righteous." This is interpreted to mean that to expound "the work of creation (*ma'aseh bereshit*)" is prideful and dishonors god; anyone who does so should be silenced. That "the work of creation" should not be openly discussed is a tradition that goes back to the Mishnah and Tosefta[106] and is also attested by Origen. The latter reports that the rabbis of his time reserved the teaching of four texts for mature students only: "the beginnings of Genesis, where the creation of the world is described; the beginning of the prophecy of Ezekiel, where the doctrine of the cherubim is expounded; the end [of Ezekiel], which contains the description of the future time; and this book of the Song of Songs" (Commentary on the Song of Songs, prologue).[107]

Curiously, *Genesis Rabbah* invokes the interdiction against publicly divulging the mysteries of creation, only seemingly to violate it flagrantly in the same breath. Once again, the analogy of a king building a palace is invoked:

> Rab said, . . . "Under ordinary circumstances, if a mortal king builds a palace in a place where there had been sewers, garbage, and junk, will not whoever may come and say, 'This palace is built on a place where there were sewers, garbage and junk,' give offense? So too, will not whoever comes and says, 'This world was created out of chaos, emptiness, and darkness' give offense?" R. Huna in the name of Bar Qappara: "Were the matter not explicitly written in Scripture, it

> would not be possible to state it at all: 'God created heaven and earth'—from what? From the following: 'And the earth was chaos.'" (Genesis Rabbah 1.5.1)

This time, the comparison is calculated to accentuate the potentially offensive character of the scriptural account of creation. What kind of god would build his cosmic palace in a dump? The "lying lips" of one who makes such an insulting suggestion should be rendered mute. Yet, as Rab points out and Bar Qappara confirms, the insult is inscribed in *torah* itself, which renders it not only thinkable but speakable—if only just barely. How is this paradox to be resolved? The theme of secrecy continues in the next section, as Rabbi Judah bar Simon applies Daniel 2:22 ("He reveals deep and secret things") to Genesis 1:1, noting that "the matter was not spelled out" in the first verse of the Bible: the hermeneutical mysteries must be teased out of other passages (1.6:4). Yet the mystery of the scandalous "chaos, emptiness, and darkness" remains obscure.

Slightly later in the chapter, the topic of things that are present before creation returns. "Six things came before the creation of the world," we are told: *torah* and the divine throne were actually created before the cosmos (a position consistent with the notion that *torah* is the blueprint for the cosmos), while the patriarchs, Israel, the temple, and the name of the Messiah were merely contemplated, or conceived mentally (1.4.1). But this is not the commentary's only version of the "six things" that come before the cosmos. The next section declares that god, like any builder, "requires six things: water, dust, wood, stones, canes, and iron"; however, "the *torah* came before those six things" (1.8.1). Still another version of the six is subsequently presented, this one in the form of a polemical encounter:

> A philosopher asked Rabban Gamaliel, saying to him, "Your god was indeed a great artist, but he had good materials to help him." He said to him, "What are they?" He said to him, "The unformed, void, darkness, water, wind, and the deep." He said to him, "May the spirit of that man burst! All of them are explicitly described as having been created by him. The unformed and void: 'I make peace and create evil' (Isa. 45:7). Darkness: 'I form light and create darkness' (Isa. 45:7). Water: 'Praise him, you heavens of heavens, and you waters that are above the heavens' (Ps. 148:4). Why? 'For he commanded and they were created" (Ps. 148:5). Wind: 'for lo, he who forms the mountains

> creates the wind' (Amos 4:13). The depths: 'When there were no depths, I was brought forth' (Prov. 8:24)." (Genesis Rabbah 1.9.1)

This striking exchange brings us once again to the second verse of Genesis, which yields such a troubling interpretation in the second section of the chapter. Three things—"chaos, emptiness, and darkness"—have now become six, with the extension into the second half of the biblical verse. Moreover, it is a philosopher rather than a rabbinic interpreter who proposes the comparison of god to an artist or architect using preexistent materials, and this time the rabbinic interpreter, Gamaliel, responds with notable hostility, cursing the philosopher for insinuating that god is dependent on materials that he has not made. What sense are we to make of these juxtaposed interpretations?

Immediately following the exchange between the philosopher and Gamaliel is a discussion of why Scripture begins with the Hebrew letter *bet* rather than *alef*, which returns us to the theme of secrecy: "Just as the letter *bet* [ב] is closed [at the back and sides but] open in front, so you have no right to expound concerning what is above or below, before or afterward" (1.10.1). Heaven, hell, the "work of creation," the eschaton—none of these should be openly discussed.[108] The form of the letter *bet* is a reminder that some matters must remain hidden: we should only look to the future immediately in front of us. Apparently not fully satisfied, however, the commentator repeats the query: why was the world created with a *bet*? Another reason is offered: "Because that is the letter that begins the word for blessing." Why not with an *alef*? "Because this is the first letter of the Hebrew word for curse." Were the text to begin with the curse of *alef*, heretics might ask, "But how can the world endure when it has been created with a word meaning curse?" (1.10.3–4). An earlier question echoes through this one: How can the world endure when it has been built on the equivalent of "sewers, garbage, and junk"? (1.5.1). This question would itself be met with a curse, had Scripture itself not suggested it, for surely the image of a world grounded in a cesspool is potentially far more compromising than the proposal that God is a skilled artist working with superior pigments.

Like Augustine and unlike Athanasius, chapter 1 of *Genesis Rabbah* thus engages the ambiguity of the second verse of Genesis. The rabbinic text repeatedly proposes that beginnings—or rather, what is already under way when beginnings begin—must remain veiled in mystery. At the same time, it tweaks the veil: beginnings *will* be discussed, however circuitously and inconclusively. We have seen that Gamaliel's curse performatively negates the philosopher's interpretation of the second verse as implying creation out of uncreated matter. Gamaliel's own counter-exegesis, however, does not simply replace this ambiguous verse with a clarifying interpretation but rather displaces the one verse with many, in a move that arguably intensifies ambiguity: as Niehoff points out, his proof texts are not uniformly persuasive.[109] Moreover, the literary context of the midrash encourages the reader to consider the possibility that Gamaliel is hiding as much as he is revealing. His divinely created "six things" succeed two alternate sets of six, positioned ambiguously as having been created before creation: this discourse of first things seems to rest on distinctly shifty foundations. (As a closer look at the text would make clear, even the number six is not altogether steady.) In addition, the demands for secrecy that bracket the public encounter with the philosopher raise questions about Gamaliel's transparency while hinting at revelations nearly unthinkable.

Privileging Gamaliel's rejection of the concept of uncreated matter, Niehoff has suggested that the text, in its fifth-century redaction, reflects accommodation to the Christian doctrine of *creatio ex nihilo*; persisting inconsistencies reflect the layering of historically distinct doctrines and the degree to which "rabbinic literature remained polyphonic within certain limits."[110] Schäfer has argued even more strongly for the coherence of the redacted commentary, suggesting that the editor works "in a deliberate and sophisticated way" to establish "two options"—creation from *torah* and creation from preexisting matter—in the first two sections from the chapter "as a kind of thesis and antithesis . . . and arguing against the antithesis in all the subsequent subunits, with a final showdown in 1:9," that is, the account of Gamaliel and the philosopher.[111] But it seems to me possible to read the text, at the redactional level, as both more unified than Niehoff deems it and more polyphonic than Schäfer deems it. Encrypted in the first chapter of *Genesis Rabbah*, I am suggesting, is the ambivalent and unsettling possibility that already in the beginning, there is "chaos, emptiness, and darkness"; there are a negation and a curse. But these truths can scarcely be uttered. How, indeed, does the world endure?

Reimund Leicht notes that the concerns of this first chapter are less cosmological than theological, having to do with the honor of the creator; more strictly cosmological concerns are taken up later in the text.[112] However, here as elsewhere in late ancient discussions of divine creation we see that cosmology is often entangled with theology. To acknowledge the frailty of creation is also to risk attributing weakness to the creator and vice versa. According to the rabbinic commentators, the divine *torah* writes the universe as formlessness, void, darkness, water, wind, and abyss; it writes the universe as a palace built on a cesspool. It writes even what seems scandalous, even what disgusts—even what *seems* to dishonor the creator. On the one hand, all life produces waste, and what is waste to one life-form is potentially nurture to another. There is no shame in that, though there is danger in ignoring its necessary balances. On the other hand, human "palaces," writ large and small, produce "sewers, garbage, and junk" on a shameful scale; and then we all find ourselves living on top of it, so to speak. If this is what *torah* writes, then it has also written a kind of freedom into the universe—for better and worse. Perhaps there is no shame in that either.

Interlude: Beginning Again with Scripture

Beginnings are glimpsed in a text. We call that text Genesis, from the Greek: *creation*. Or, with Plato: *becoming*. *Khora* is "the receptacle, the nurse, as it were, of all genesis" (Timaeus 49a).

Can we hear the verb echoing through the noun? Genesis names a thing—*the all*, we might call it—and also an ongoing natural-cultural process—creating, making, birthing, becoming.

Genesis is the book of creating, of making, of birthing, of becoming. It is the book of everything that has a beginning. It is the book of everything, then—of all that was or is or will be.

Only beginning has no beginning: it is always already under way. Genesis is the book of beginning as well.

Beginnings are glimpsed in Genesis, the text that sits at the beginning of the Bible—the text that begins the Bible with an account of beginnings always already under way.

Genesis does not merely describe, then. It acts; it performs. It begins, creates, makes, births, becomes. It writes the universe. It is the nurse, the matrix of all becoming. It is the gateway to what Jorge Luis Borges calls the "Library of Babel," a cosmos of writing that is infinitely expansive, containing all possible books.

To read Genesis is also to write and rewrite genesis, to become part of the writing of the universe. It is to cocreate. The question is not how to begin: it has already begun. The question is what kind of world we want to create together.

In/Conclusion: *Khora*, God, Materiality

Khora, God

Timaeus's speech begins with the barely believable story of a demiurgical deity. It begins again with the barely comprehensible account of the mysterious *khora*. *Khora* and the god do not occupy the same discursive space: the god exits when *khora* enters, and *khora* disappears when the god returns. But they do somehow need each other. We might say that the fiction of the god is the explanatory machine, the irrational nonsense of *khora* the event that the machine produces and that also jams the machine. Timaeus's speech suggests that there is no other way to approach an understanding of the elusive world of becoming.

There is another sense in which *khora* and the god need each other. The dialogue does not present a single, unified model of genesis. It wavers "between a discourse of production and a discourse of procreation and birth," as Sallis notes.[113] The god is presented as a demiurge or an artisan, but is nonetheless also described as a father. *Khora* is presented as a maternal or nurturing figure, but nonetheless also plays a crucial role in the god's crafting of the world. The sustained dichotomy of god and *khora*, *techne* and genesis, culture and nature, insinuates a model of creativity that is not reducible to either building or birthing and yet can only be described by analogy to each.

Ancient Jewish and Christian theologians inherit the understanding of god as both "maker and father" of the cosmos (Timaeus 28c), while generally ignoring the figure of *khora* as such, as we have seen. Leaning into the demiurgic model, they must walk a fine line, claiming that their god is both like and unlike an artisan. To the extent that he *can* be compared to an artisan, he is one who makes his own materials; and those materials are not, then, simply . . . materials. They are—or rather, *it is*—something much stranger: no-thing, nearly nothing, nothing-something, formlessness, void, darkness, abyss. It has no name and many names: we have called it *khora,* even if they did not.

The same theologians also lean into the erotic model of generation, though the metaphors are partly transposed. The god creates by speaking, conceiving, or writing the universe, they suggest. Here too we discern traces of *khora*, always already present with the divine *logos*, *sophia*, or *torah* that generates the world. It is there as the very possibility of materialization, when word is voiced, wisdom delivered, scripture inscribed on the page. It is there when the god does not merely speak, conceive, or write the universe but also creates space for a response—when the god *needs* a response. *Khora* is there, at the heart of a universe ever emerging in the free play of call and response. It is there as radical openness; as a dispersal of affectivity, at once receptive and rogue; as a capacity for ongoing formation, deformation, reformation. The god is there, at the heart of the universe, too. But it is not the transcendent, all-powerful god we began with when we first told the story. It is not the divinely determined universe we began with either.

The doctrine of creation ex nihilo that emerges in late antiquity—and it is not really a single doctrine, as we have seen—does not, then, eliminate the khoric. Paradoxically, it seems to reintroduce it. Nor does it eliminate divine finitude. Paradoxically, it seems to require it.

The traces of the ancient cosmotheology that I am describing may be seen to anticipate certain movements within contemporary thought, occurring on the borders of process theology and continental philosophy. Richard Kearney proposes a god "who may be," wagering "that *khora* is neither identical with God nor incompatible with God but marks an open site where the divine may dwell and heal."[114] John Caputo argues in turn for a "weak" god, a god of "perhaps"; at the same time, he suggests that "without *khora* there is no 'perhaps.' *Khora* is . . . the slash between maybe/maybe not."[115] Taking his lead from Catherine Keller, Caputo discovers khoric affinities in the figure of *tehom*, the abyss or deep of Genesis 1:2. "There is an element of irreducible indeterminacy and instability built right into creation," he suggests, "so that creation is going to be continually exposed to re-creation."[116] However, Caputo insists provocatively that "we must avoid speaking well of *khora*, avoid praising (and praying) to her/it as the giver of all good gifts, as the generous mother who engenders all."[117] Unlike god, *khora* does not call; *khora* does not give; *khora* "could care less."[118]

In partial contrast to Caputo's austerely Derridean chorology, Keller's musings on *khora* in the guise of *tehom* draw upon a rich range of biblical figures and theological traditions. Calling for "an apophatic panentheism," Keller suggests that "the godhead, or rather the godness, *'in' whom unfolds* the universe can be theologized as Tehom," while "the divinity *who unfolds 'in'* the all is called by such biblical names as Elohim, Sophia, Logos, Christ."[119] A third, Spirit—"the relationality itself"[120]—completes her trinity: chaos, difference, relation. Here *khora* is not in opposition to god, not an alternative to god, not even a complement to god. Rather, it is an aspect of divinity itself—"the depth of god," as Keller puts it.[121] This triadic god is immanent in creation, yet it is not simply identical with the universe in which it is unfolded and which it enfolds.

It is not my purpose to suggest that any of these contemporary theologies, here so briefly sketched, aligns itself precisely with any of the late ancient theologies that we have been considering. The resonances may be illumining, however. Moreover, they complicate certain assumptions regarding the implications of creation ex nihilo for both theology and cosmology. As Mary-Jane Rubenstein puts it, "The scholars who engage the *ex nihilo* tend to stake powerful claims either with or against the doctrine." She notes that proponents as well as opponents of the doctrine agree that it "affirms the absolute sovereignty of the creator."[122] Indeed, both Keller and Caputo distance themselves from the doctrine in the strongest possible terms, refusing divine sovereignty and omnipotence in the name of a theology of divine-creaturely cooperation.[123] Yet I question whether the ex nihilo secures divine sovereignty—correspondingly, creaturely passivity—as successfully as is generally supposed. Part of the burden of this chapter has been to show that it does not. Sallis suggests that "the chorology [of the *Timaeus*] could be said to bring both the founding of metaphysics and its displacement, both at once. Originating metaphysics would have been exposing it to the abyss, to the abysmal χώρα, which is both origin and abyss, both at the same time."[124] Might we not say much the same for metaphysics' double, ontotheology, in the guise of the ex nihilo? Divine creation out of nothing turns out to be both the origin and the abyss of a theology of divine sovereignty, I am suggesting. The same gesture that banishes *khora* also summons it—inadvertently, so to speak. And as long as *khora* is part of the picture, it challenges the fiction of a god of unfettered power who simply wills the world into existence and who continues (for better or worse) to subject that world to his will.

Keller herself stresses that the khoric can never be absolutely eliminated, however much theologians may try. Proposing that ex nihilo theologies have "systematically and symbolically sought to erase the chaos of creation," she adds that "such a maneuver . . . was always doomed to a vicious circle: the nothingness invariably returns with the face of the feared chaos—to be nihilated all the more violently."[125] In addition to such violent cycles of repression and the return of the repressed, something more promising also takes place, I am suggesting—something more like a philosophical auto-deconstruction. Within accounts of divine creation out of nothing—or out of a divinely generated, precosmic *almost* something (the slipperiness here is instructive)—the duality of god as being and creation as (out of) nonbeing does not quite hold. As cosmic word, wisdom, or text—as *beginning* itself—the god evokes orderings of differentiation and interrelation. But it does so in relation to an otherness that is always already present, in every beginning, whether by necessity or by divine will—or by divinely willed necessity. That otherness is the possibility of possibility; it is flow, mutability, unpredictable change; it is radical receptivity; it is contention, disorder, disintegration; it is the ultimate elusiveness of all things, their spooky mystery. It has no name and many names, as we have seen. We may call it *khora*; we may call it divine, or not. But without the unknowing of the third, knowledge of both god and the world will yield only false certainties.

Khora, Materiality

What more can we say about *khora* and materiality? We have seen that Aristotle is seemingly the first to identify the Platonic figure with *hyle*, or matter, or rather with what he calls "primary matter"—that which "underlies" formed bodies, or "substances" (On Generation 329a). Yet, as we have also seen, for Plato the relation of *khora* to materiality is more complex than this formulation suggests. Itself imperceptible to the senses, *khora* is that "in which" the materialization of forms takes place (Timaeus 50c). It is what enables things to "cling to being at least in a certain way, on pain of being nothing at all" (52c). Dubbed the "nurse of becoming," it can be compared to a winnowing basket filled with fluid and mobile elemental potencies that cause it to shake and reel; its own movement moves them in turn (52d–e). The Platonic *khora* is not, then, matter in the sense of passive, inert stuff but rather the very power and possibility of materialization; it enables—or perhaps better, necessitates—the liveliness, agency, and fluidity of fleshly existence, as well as its transience and tenuousness.

Later Jewish and Christian thinkers inherit the ambiguous conflation of *khora* and *hyle*. They also inherit a suggestive biblical text, abstracted from more ancient myth: "In the beginning, god made the heavens and the earth. And the earth was invisible and unformed, and darkness was over the abyss, and a divine wind was being carried along over the water" (Gen. 1:1–2, LXX). Since these verses are understood to describe events taking place before the creation of the visible cosmos, many late ancient interpreters are inclined to view the "heaven and earth" of verse 1 and the mysterious entities of verse 2 as precosmic elements of some sort. Whether or not divinely made (as most Christian readers assume they are), they do seem to be there "in the beginning"—a reading reinforced by the demiurgic model inherited from the *Timaeus*. At the same time, as we have seen, precosmic elements are not generally interpreted as matter per se but as something more equivocal, mysterious, even scandalous—something more khoric, in a word.

Chris Boesel and Catherine Keller propose the phrase "apophatic body" to express the sense of "an unspeakable body" that eludes "mastery," whether "theological, political, or epistemological." As a gesture of negation, apophasis "implies the unsaying of the impermeable boundaries and ontological—indeed ontotheological—fixities of a modern body."[126] This conceptual move, which builds on traditions of apophatic or "negative" theology, participates in a broader emergence of "new materialisms." Theorists working across a range of disciplines push for an "ontological reorientation" that "conceives of matter itself as lively or as exhibiting agency"; they emphasize "materialization as a complex, pluralistic, relatively open process" in which humans are "thoroughly immersed."[127]

If I have argued that there are traces of such a "dark cosmology" in our ancient texts, human exceptionalism is arguably a sticking point, and no small one. There is much in these writings that tugs against a view of humans as "thoroughly immersed" in material processes, or as just one among many in a "democracy of objects," as Levi Bryant names it.[128] Yet one may detect other sensibilities at work as well. Our authors break with the *Timaeus* to insist that one god creates all things in the same way, without mediation: the god is every bit as much the creator of the frog as of the human. They insist as well that the beauty and goodness of the universe lie in its interconnected totality. They imagine themselves citizens not of one city but of the very cosmos. Finally, exceptionalism itself is a double-edged sword when humans are deemed exceptional precisely in their capacity to sin by falling away from the divine *logos* that binds the universe in a harmonious whole. Is that not also the lesson of the "anthropocene"—that humans bring ruin to creation precisely by denying their connection to all things? "We are becoming aware of the world at the precise moment at which we are 'destroying' it—or at any rate, globally reshaping it," as Morton puts it. He adds: "We have barely become conscious that we have been terraforming the Earth all along. Now we have the chance to face up to this fact and to our coexistence with all beings."[129]

Khora, Materiality, God—Beginning Yet Again

Timaeus confesses that speaking of *khora* is "difficult and obscure" (Timaeus 49a). It can only be known "by a sort of bastard reckoning, hardly trustworthy" (52b). *Khora* unsettles knowledge, then. It also unsettles feelings. Reading Genesis as well as Plato's *Timaeus*, Philo encounters the bleakness of a "yawning and desolate and empty *khora*." Origen discovers the universe to be "a vast and monstrous animal," while Athanasius recoils from the abyss of nonbeing from which all things emerge. When Augustine attempts to imagine the mutability of materiality, he is seemingly disturbed and fascinated in equal measure. The rabbis shudder at the thought of a universe built on the equivalent of "sewers, garbage, and junk"—but they hold onto the thought nonetheless. As Morton comments, "We can't spit out the disgusting real of ecological enmeshment. It's just too close and too painful for comfort. So it's a weird, perverse aesthetics that includes the ugly and the horrifying, embracing the monster."[130]

How does one stay steady in the face of what is deeply unsettling? See beauty in the wastelands? Love what horrifies? These are questions that have always pressed, or should have. They press with more urgency than ever, as we begin to experience the realities of ecological devastation that has tilted past the tipping point. Here we have been exploring one possible answer to such challenging questions: we stay steady, we see beauty, we love by continuing to tune into the generative potencies of what Timaeus calls *khora* and also by continuing to tune into the divine. If our ancient texts have yielded eco-chorologies, they have also been eco-theologies. But why begin again with theology *now*?

Not all of us will want to, to be sure—and for good reasons.

Some of us will. But to what end?

I take my own cue from the radical theologians of *khora* already mentioned—Keller, Caputo, Kearney—as well as Sharon Betcher. In particular, Keller and Betcher write directly into our planetary moment of ecological precariousness. Neither takes the relevance of talk of god for granted, but neither declares it irrelevant either—on the contrary. As we have already seen,

Keller's evocation of god as the self-transcending capacity of the world embraces both chaos and differential relationality—both *khora* and *logos*, in other words—as well as the relation between them, her trinitarian third. Following Nicholas of Cusa, she suggests that this triadic god names *possibility*—the possibility that emerges at the borders of the possible and the impossible, the knowable and the unknowable, when (for example) "it is still *possible* to avert unspeakable disaster, yet improbable that we will."[131] God also names "a love-relation to your widest world," as she puts it.[132] God is a *doing* as much as a name or a naming, operative in the movements of "collective materialization" that channel our passionate love, "binding us to one another while stimulating our singular gifts."[133] To speak the name of god—to participate in the doing of god—is to risk both hope and love, then. It is to respond in a particular way to the question, "How shall we greet the unknown before us?"[134] Keller writes, "Across the threshold of catastrophe, the convivial cosmopolis can—posse ipsum—yet coalesce. There is no God-guarantee on the outcome; but there is the lure."[135]

Betcher shares Keller's conviction that theology has something distinctive and significant to communicate in this moment of planetary fragility and uncertainty, yet she distances herself even more than Keller from traditional theological discourse. In so doing, she addresses herself directly to the cultural reality that it has become increasingly difficult for many Western Christians (among others) to profess to a belief in god. Where god may no longer easily be called upon, spirituality may be more relevant than ever, however. Replacing language of "god" with that of "spirit," Betcher suggests that precisely amid the ruins of our postapocalyptic world, "spirit" may be "a necessary 'prosthesis,' an aid that might help us advance toward spacious and fearless empathy, toward forbearance amidst messy entanglements."[136] The term *prosthesis* does double work for her. On the one hand, it names a supplement in "what can augment or capacitate 'belief in the world,'" enabling "the mutual submission—or entrustment—of flesh one to another."[137] On the other hand, it evokes the figure of disablement—the "crip"—who stands in for "all the damned and damaged" of the planet, and marks the pressing challenge of learning to navigate "disgust, fear, and pain otherwise than by encultured avoidance."[138] "Keeping faith with the crip is, it seems to me, not distinct from keeping faith with a world of becoming at this epochal turning," Betcher professes.[139] Spirit is her name for that capacity for "keeping faith" through the transformation of aesthetics and affect and the generation of what she calls "corporeal generosity" amidst the precariousness of fleshly existence.[140] Spirit evokes the resilience and resourcefulness of the planetary collective located precisely in its sites of greatest vulnerability. It manifests in the choice to risk love and hope and in the opening to joy and beauty amid the ruins.

We begin again with theology, then; or some of us do. It is a form of theology in which *khora* and *logos* are brought into the most intimate relation, in and as the immanence of the divine.

But why begin again with *ancient* theology?

We might think of it, first of all, as a kind of recycling, an act of conservation. As Keller puts it, "saving" theological traditions "might be just good ecology. Why waste every metaphor of our infinite entanglement?" she asks.[141] If we are to alter our sensibilities in the dramatic ways required for thinking "the ecological thought," as Morton dubs it, we will need *all* of our best cultural resources. And who knows what might turn out to be a resource? It is impossible to say in advance. To meet a nearly unimaginable future, we may do well to bend back toward a nearly unfathomable past. To cite Keller again, "Without our widest aeon, our pasts recapitulated in the counter-narratives and queer temporalities of our most becoming perspectives, how can we greet the unknown before us?"[142]

We might also think of this bending back to antiquity, this beginning again with antiquity, as a practice of mindfulness—a contemplative slowing-down to peruse texts that speak to our own moment only indirectly. We might think of it as a cultivation of humility as well. If the anthropocene marks an era of history in which human and geological temporalities intersect—in which humans have made a geologically significant difference—at what point can that difference be said to have been inscribed? "There are a whole series of thresholds, ranging from sedentary agriculture, colonization, the steam engine, nuclear energy and capitalism," notes Claire Colebrook. (Some might add "Platonism" or "Christianity" to the list.) As Colebrook points out, "The time of politics and the time of the planet, once deemed to be distinct, are now colliding, but not converging."[143] Indeed, from the perspective of human time, the anthropocene might be said to have been long in the making. But might the same not be said for forms of human thought and practice that acknowledge that "everything is interconnected"?[144] Our own most novel thoughts are themselves complexly interconnected with past ones—even, or especially, with past ones we wish to repudiate. That there might be traces of an eco-chorology inscribed in the very beginning of ontotheology—could that be modest cause for hope?

Postlude: Beginnings, Again

In the beginning. . .

What is a *beginning*?

It is a speaking, a conception, an inscription. A child. A teacher. A secret. It has so many names! It might be a blessing. It might be a curse. Anything is possible, in beginning. Anything and everything is present in a beginning.

In the beginning, the god. . .

What is the *god*?

To know the maker and parent of this universe is most difficult—no, it is impossible. But we may say nonetheless that the god creates. (Call it a fiction or not.) The god incites love, hope, joy. The god incarnates—materializes. The god is strong; it is the name of what is strong in the world. And the god also needs help—a response.

In the beginning, the god was creating heaven and earth.

What is *heaven and earth*?

It is complicated, so complicated!

It is full of difference, discord, and incoherence. It is also the ever-emergent harmony of interconnected parts.

It is a yawning and desolate and empty *khora*. It is a cesspool and what emerges from that. It is also the plenitude of an elemental flow.

It is vulnerability and destructiveness. It is also a birthing mother or a gushing breast.

It is a vast and monstrous animal. It is things becoming other things. They are all the same and all different.

It is no-thing. It is nothing-something. It is everything and more than everything.

It makes me shudder with horror and burn with love.

II

Queering Creation

Hagiography Without Humans

Prelude: Ecocriticism as Queer Theory

> Ecology is the latest in a series of humiliations of the human. . . . Ecological humiliation spawns a politicized intimacy with other beings. . . . Such intimacy necessitates thinking and practicing weakness rather than mastery, fragmentariness rather than holism, and deconstructive tentativeness rather than aggressive assertion, multiplying differences, growing up through the concrete of reification. . . . To contemplate ecology's unfathomable intimacies is to imagine pleasures that are not heteronormative, not genital, not geared to ideologies about where the body stops and starts.
>
> —Timothy Morton

In a 2010 article titled "Queer Ecology," ecocritic Timothy Morton predicts that a meeting between ecological criticism and queer theory would produce a "fantastic explosion";[1] outlining the reasons why, he goes a long way toward staging such a meeting. Nor is Morton the only one to have been intrigued by the potential of this particular convergence: the years 2010 and 2011 also saw the publication of a multiauthor volume titled *Queer Ecologies: Sex, Nature, Politics, Desire*,[2] as well as the appearance of both a popularizing article—Alex Johnson's "How to Queer Ecology"[3]—and a scholarly critique of the very project—Greg Garrard's "How Queer Is Green?"[4] Subsequent scholarship has continued to build on the insights that emerge at the convergence of ecological and queer theories, adding animal studies and disability theory to the mix.[5]

These recent meetings of lavender and green scholarship[6] emerge against the backdrop of a longer ecofeminist conversation. Among the significant forerunners is Greta Gaard's 1997 article "Toward a Queer Ecofeminism." Building on the coalitional politics and intersectional analysis of third-wave feminism, Gaard homes in on the resonance of ecofeminist and queer theories at two critical points. First, both ecofeminists and queer theorists interrogate the sexualization of "nature" and the naturalization of "sexuality." Second, both ecofeminists and queer theorists expose the particular role played by Christianity in the "erotophobic" oppression (as Gaard names it) of women, queers, and nature.[7] One is left with the distinct impression that if queer ecofeminism met Christian theology, there would be an explosion of an altogether different and less pleasant sort than the one Morton images for ecological criticism and queer theory—a point I shall hope to complicate.

As Gaard already intimates, ecological criticism and queer theory share a critique of the concept of nature, typically embedded in a nature-culture binary. This binary suggests that humans—or, rather, some humans—occupy the privileged space of "culture," in relation to which "nature" is foreign and other. Nature is "always 'over yonder,' alien and alienated," as Morton puts it.[8] Commonly aligned with the abjected realms of the fleshly and the erotic, nature is, at best, to be used and managed, saved and conserved, but always also to be dominated, controlled, and tamed—even to the point of eradication. Following a different logic, alien nature is not abjected but idealized—"a self-contained form suspended afar, shimmering and naked behind glass like an expensive painting."[9] As ideal, nature lends its authority to certain subjects at the expense of others: to the heterosexual at the expense of the homosexual, for example, or the healthy at the expense of the diseased or disabled.[10]

Resisting this problematic binary, many contemporary ecological thinkers follow Donna Haraway in invoking the term "naturecultures" to indicate the inextricability, perhaps indistinguishability, of what we typically think of as separate or even opposed domains.[11] Others advocate dispensing with the concept of nature altogether:[12] there is no "nature" as such; there are only multiple, diverse natures, and humans are part of the mix. Moreover, these natures are neither altogether distinct nor completely stable. In Morton's phrasing, "life-forms constitute a *mesh*, a nontotalizable, open-ended concatenation of interrelations that blur and confound boundaries at practically any level: between species, between the living and the nonliving, between organism and environment."[13] The diversity and fluidity of life-forms is matched by a diversity and fluidity of sexes and sexualities. These include a variety of asexual modes of reproduction as well as ones involving multiple genders. Same-sex sexual activities have been documented in a wide variety of species, from dragonflies to lizards to penguins to giraffes to humans[14]—though that is not necessarily what makes living things queerest. "What we call Nature is monstrous and mutating," declares Morton, "strangely strange all the way down and all the way through."[15]

Such thinking is distinctly at odds with the ethos and ideology that the environmental movement has historically fostered. Beginning in the mid- to late nineteenth century, anxieties about masculinity focused on the city as a space of artificiality, pathology, and effete culture—a breeding ground for homosexuality. Rural and wilderness spaces, in contrast, were seen as natural, healthy, and ruggedly masculine and heterosexual. "Parks were a curative response" to the ills of the city, as Bruce Erickson and Catriona Mortimer-Sandilands note: "They were created in part as places in which heterosexual masculinity could be performed and solidified . . . through rigorous, health-giving recreation."[16] Or as Morton depicts the mind-set: "Rugged, bleak, masculine Nature defines itself through contrasts. It's outdoorsy, not 'shut in.' It's extraverted, not introverted. It's heterosexual, not homosexual. It's able-bodied—'disability' is nowhere to be seen."[17]

A "muscular" Christianity has frequently colluded with such a masculinist, heterosexist, and ableist environmentalism, in the nineteenth century and well beyond. But does the real problem with Christianity not originate in a much more distant past? Here we return to Gaard's critique: Christian "erotophobia" can indeed be traced to Christianity's ancient beginnings. Yet I want to suggest that those beginnings also harbor other legacies. As Mayra Rivera has recently argued, there are at least two distinct Christian understandings of flesh that emerge in antiquity. If phobia marks the Pauline tradition, which fantasizes pristine spiritual bodies cutting loose from the messy entanglement of flesh, the incarnational tradition of the Gospel of John embraces flesh as the site of salvific transformation.[18] What Rivera marks as a Johannine view resurfaces in later antiquity in the queerly exuberant eroticism of hagiographical literature, as I have argued elsewhere.[19] It is the task of the following pages to explore late ancient saints' *Lives* as possible narrative resources for thinking forward in both a queer and a queerly ecological mode.

Hagiography bends and blends the genres of romance and biography that emerged in the Hellenistic and early Roman imperial periods. In so doing, it queers time, place, and finally the human. *It puts the human as such into question.* Time warps when the purposeful arc of a life centered on production and reproduction is arrested or interrupted, redirected or dissipated, stretched out or condensed. Place is likewise transformed as distinctions between city and wilderness dissolve and the local opens up onto the global. The desert becomes a city, as Athanasius famously asserts (Life of Antony 14); is the city then also a kind of desert? When militant monks flee to uninhabited regions to prove themselves in battle with demons, they do not return to civilization like so many triumphant weekend warriors, identity confirmed, masculinity honed. Rather, the desert humbles them; it undoes them, tossing up specular hallucinations that blur the line separating males from females or humans from other creatures. In the process, desire is reoriented and also intensified, as sexuality is dislodged from the human institutions of marriage, slavery, and prostitution that traditionally defined and confined it. Ascetic saints are both subject and object of that reoriented desire, or, rather, the saint is the site where the distance between subject and object threatens to collapse. "Strange stranger" par excellence,[20] at once foreign and familiar, repelling and alluring, the saint shows us the world as an open and unbounded web of intimate aliens, of which we ourselves are a part. The saint reminds us of the porosity, interpenetrability, and precarity of all flesh, and also of its mysterious allure. Incarnating a promiscuous "biophilia," as Edward O. Wilson names it, the saint exposes "the innately emotional affiliation of human beings to other living organisms."[21]

The premodern saint I am summoning is, most obviously, a figure of literary representation. Yet I want to attend to the ways that the saint also exceeds the bounds of literary representation, arriving in and as an assemblage of textual, visual, and ritual productions better understood as performative than as representational. Even a literary text may have a strong performative dimension, especially when it functions within a larger performative collectivity. From this perspective, the literary *vita* does not merely depict the saint through an image or likeness: rather, it makes the saint, so to speak; the same might be said of an icon or a liturgy.

From another perspective, text, icon, and liturgy all mediate the saint's own performance, even as the saint emerges as the unstable effect of such media. That performance is a self-performance, a *techne*, or technique, of self-(de)formation that is also a spectacle of self-display—a kind of performance art, then. And like much recent performance art, the saint's act disturbs and challenges the spectator with its insistence on the irreducible corporality—the abysmal corruptibility—of human being. At the same time, it uncovers in those very abject depths a transcendent beauty, all the more powerful because flawed and transient. This sensibility is akin to what Tobin Siebers refers to as a "disability aesthetics" that embraces the human as fragmented, broken, or wounded, thereby shifting the terms of what counts as human and also what counts as art.[22] It is likewise akin to the sensibility evoked by what Morton calls "art in the age of asymmetry"—that is, art of our own time in which the medium and the message mutually outstrip one another. Producing excesses of both materiality and spirit that do not simply balance each other out, as classical works are understood to do, such art gives rise to a postmodern animism—in Morton's terms, "animism *sous rature*." This is an animism attuned to the alluring mystery and uncanny liveliness of more-than-human beings and to the often uncomfortable intimacy of "coexistence without center and edge."[23]

In pursuit of the saints of Christian hagiography, we begin by tracking one who precedes them, one they themselves follow—namely, the saintly philosopher Plotinus. In the biography written by his disciple Porphyry, we encounter the holy man in an unsettling display of excessive animality coupled with uncanny transcendence—what Derrida calls divinanimality.[24] We encounter a life that breaks with the scripted chronology of a human *bios*, opening onto the queer temporality of *zoe*, the vitality of flesh and spirit that traverses bodies and species. Finally, we encounter a life that teaches readers to desire differently. In all these ways, Plotinus prepares us for the slightly later literary performances of Christian holy men and women. The *Lives* of Antony, Paul, Mary of Egypt, all intertextually linked, allow us to consider how hagiography as a genre queers time, place, desire, and the human, thereby lending itself to ecological thought. The *Life of Syncletica* in turn invites a reading of the saint as embodying a vivid aesthetics of disability, while the *Life of Simeon* invites a reading of the saint as a complex posthuman assemblage. These premodern saints may help teach us how better to live into our own anthropocene "age of asymmetry," in which human knowledge of the universe is both radically expanded and severely humbled. They may show us what it might look like to practice "weakness rather than mastery, fragmentariness rather than holism," and to embrace "ecology's unfathomable intimacies."[25]

Before Hagiography, Autozoography: The *Life of Plotinus*

In a now well-known lecture, *L'animal que donc je suis* (*The Animal That Therefore I Am*), Jacques Derrida issues a challenge to the particular humanism conveyed by the Cartesian dictum *je pense donc je suis*. His linguistic pivot point is the word *suis*, the first-person singular form of both the verb "to be" (*être*) and the verb "to follow" (*suivre*). Making the most of this double meaning—*I am, I am following*—Derrida displaces the subject from the firm bedrock of *being* onto the shifting sands of *following*, while at the same time rendering *following* an elusive signifier of the complex human relation to *animality*. Where modern philosophy has defined "the animal" as that which is excluded by the quintessentially human autobiographical performative "I am," Derrida discovers in the "I follow" a challenge to both the autonomy and the stability of the subject and hence to its exclusive humanity. Pursuing a thicker, richer conceptualization of difference, he also seeks a more animalistic autobiography—or perhaps better, autozoography. Seeming to enact such a possibility, he depicts himself standing naked before the enigmatic and unsettling gaze of a particular cat, struck by an inexplicable sense of shame. "Who am I . . . ? Who is it that I am (following)? Whom should this be asked of if not of the other? And perhaps of the cat itself?" he queries.[26] "And say the animal responded?"[27] Derrida continues to track the implications of the ever-more-ambiguous *je suis*, announcing that "everything in what I am about to say will lead back to the question of what 'to follow' or 'to pursue' means, as well as 'to be after.'"[28] The possibilities multiply: "Being-after-it in the sense of the hunt, taming, training, or being-after-it in the sense of succession or inheritance?" he wonders.[29] "There is the question of following, of the persecution or the seduction of the other," he insists.[30] Haunted by the threat of violence, the *je suis* takes the form of a confession: *I am pursuing, chasing, tracking, seducing, persecuting, coming after* the other, animal; this is also, however, to say, I *am* the other animal. The relations of following are, moreover, layered and reversible. "Who comes before and who is after whom?" he asks.[31]

Derrida observes that the third-century philosopher Porphyry is among those pre-Cartesians who insist on the capacity of animals to respond to a call, a responsivity that constitutes a primal autobiographical gesture, as if to say, "Here I am." He cites Porphyry's reference in his treatise *On Abstinence* to "a lamprey which belonged to the Roman Crassus [which] would come to him when called by name, and had such an effect on him that he mourned when it died."[32] Here I want to track the autobiographical voice of Porphyry himself, focusing on his *Life of Plotinus*. There being and following converge in the identity of the philosopher: Porphyry is following the other, philosopher; Porphyry is, therefore, the other philosopher. "Hunt, taming, and training" are as relevant as "succession or inheritance," and the disciple's pursuit of the teacher is also staged as a forceful seduction. Porphyry persuades Plotinus to write when he is more inclined to converse; he also dominates and domesticates his master's untamed texts. But who is after whom? The seducer is always already seduced: Plotinus is pursuing Porphyry as well, not least by withdrawing from him, leading him along deliberately digressive paths of thought.

If Porphyry follows Plotinus, he comes after and is caught in the enigmatic and unsettling gaze of a particular philosophical animal. To be precise, he is caught in the enigmatic and unsettling gaze of a philosophical animal who has poor eyesight, questionable hygiene, and constipated bowels, one who appears to be ashamed of his very embodiment and yet is strangely shameless, scarcely civilized, still preferring his nurse's breasts to grammar lessons at age eight, mispronouncing and misspelling words his whole life, indeed seemingly reluctant to bother with proper scholarship at all. Porphyry is pursuing and pursued by one who, in the singular moment of his death, appears other than human, at once serpent and god, merging with "the divine in all"—or, as Derrida might have it, with the "divinanimality" that may also be configured as "the irreducible living multiplicity of mortals."[33]

The *Life of Plotinus* fascinates for many reasons. It gives us precious biographical information about an ancient philosopher—in itself an unusual circumstance. Yet, as Patricia Cox Miller demonstrates, the Plotinus of Porphyry's *Life* is ultimately elusive; Porphyry's biography is, as she phrases it, "the baffled telling of a vision," a "veil of images," "a labyrinthine tracing and a weaving together of the tracks of soul in life."[34] The text strains with the paradox inhering in the very endeavor "to integrate Plotinus's writing with the story of Plotinus's physical body: to attach the life of the body to the life of the mind," as Derek Krueger puts it.[35] Art in an age of asymmetry? Porphyry's Plotinus exhibits too much spirit and too much carnality at once.

I, Porphyry

Perhaps every biography is at least implicitly also an autobiography—the autobiography of a disciple, of one who follows or tracks the life of another, and also the autobiography of a successor, one who comes after, one who can claim to inherit. The biographer says in the same breath: *I follow **another***; and also, *therefore **I** am*. We might understand Porphyry's choice to begin his *Life of Plotinus* with a narration of his master's death in the context of this performative "I follow." It is precisely in the moment of recounting Plotinus's departure from life that Porphyry himself enters the text for the first time in the emphatic first-person voice—*ego Porphyrios*—that he will adopt intermittently throughout the text. "When [Plotinus] died, I, Porphyry [ἐγὼ . . . ὁ

Πορφύριος] happened to be staying at Lilybaeum, Amelius at Apamea in Syria, and Castricius in Rome: only Eustochius was present" (Life of Plotinus 2).[36] Porphyry's writing of Plotinus's life and of his own *je suis* follows Plotinus, then; it comes after him—postmortem. But if Porphyry's textual *je suis* succeeds Plotinus, it also pursues one who is elusive from the start. Plotinus's closest followers are not present when he dies; only the doctor Eustochius remains to witness the serpent that "slipped away into a hole in the wall" in the very instant of Plotinus's death (2). This animal will not be easy to track. Porphyry does not even know where or when he was born.

Plotinus himself once pursued a teacher, Porphyry relates. When he finally found Ammonius, he proclaimed, "This is the one I sought" (3). However, after spending eleven years with Ammonius in Alexandria, he developed a desire to seek new learning in Persia and India; he joined the army in order to see the world but only barely made it back from Mesopotamia when his commanding emperor was killed. At that point he settled not in Alexandria but in Rome. Following Ammonius leads Plotinus on an errant course, then. Those disciples who subsequently gather around him discover that following Plotinus is no easy task either: having pledged not to disclose any of his own master's teachings, for his first ten years in Rome, Plotinus does not write at all, and when he teaches, he draws out the questions of his companions so that his discourse, it is said, is "full of disorder and much nonsense" (3).

At this point Porphyry's autobiographical *ego Porphyrios* makes its entry once again: "In the tenth year of Gallienus's reign, I, Porphyry [ἐγὼ Πορφύριος] arrived from Greece" (4). As it turns out, Porphyry's is not only an autobiographical *ego* but also a bibliographical one. Fortuitously, Plotinus has finally begun to write. By the time Porphyry arrives, he has produced some twenty-one treatises over the course of a decade, but they are guarded jealously within an inner circle. Moreover, they are a bit of a mess, lacking titles (4) and sloppily drafted: Plotinus cannot bear to read over his works once they are written; his handwriting is terrible; and he has no interest in the proper spelling or spacing of words (8). Clearly, he is in need of a strong editor. Porphyry runs through the roll call of Plotinus's major disciples—Amelius, Paulinus, Eustochius, Zoticus, Zethos, Castricius, Rogatianus, Serapion—ending with the flourish of another emphatic self-reference: "He held me, Porphyry of Tyre [ἐμὲ Πορφύριον Τύριον], among his closest companions, me whom he deemed worthy to set his writings in order" (7). Porphyry has just provided a list of those writings, divided into three groups according to chronology: the twenty-one treatises written before Porphyry's arrival; the twenty-four written during Porphyry's six-year sojourn; and nine that Plotinus sent Porphyry after Porphyry left Rome. If the first group is still immature, the last group reflects waning powers; only those treatises written under Porphyry's own watch "attain the highest perfection," declares Porphyry (6). The bibliography itself will have to be revised, however. The *Life* concludes with a reordered table of contents for Porphyry's definitive edition of Plotinus's works, in which the fifty-four treatises are arranged in six groups of nine (the *Enneads*), ordered not chronologically but thematically (24).

Following Plotinus, Porphyry pursues, trains, and tames the other, philosopher; he urges him on to greater productivity; he straightens out his crooked letters, corrects and rearranges his texts. Following Plotinus, Porphyry seeks to master his master's corpus, with his assertive *je suis*, his *ego Porphyrios*. And yet at the same time, Plotinus is always ahead of him, leading him on and eluding his grasp. Two further instances of the emphatic first person self-designation are telling in this respect; both occur in the genitive absolute, *emou Porphyriou*—that is, in grammatical constructions that are both "loose" and subordinate, thus no longer so domineering. The first introduces an unexpected aspect of Porphyry's character. "When at one time I, Porphyry, had in mind to release myself from life [βίου], [Plotinus] perceived this; and coming unexpectedly to the house where I was staying and saying that this desire did not come from a rational state but from some illness of melancholy, he commanded me to depart. Being persuaded by him, I left for Sicily" (11). Porphyry here refers to his own suicidal impulse as a desire to release himself from *bios,* rather than from *zōē* or *zēn*—the more typical usage in this construction. It is not the animal life of *zōē* that he seeks to escape but the distinctly human *bios,* in other words. Perhaps that is the unsettling effect of following Plotinus for one not made of sterner stuff: compassionately, Plotinus orders him from his presence. Thus it is that Porphyry lives to tell the tale of his master's death; thus it is that he does not witness that death himself. But perhaps he has begun to revise his aims, turning from auto*bio*graphy to auto*zoo*graphy.

The second instance of emphatic self-designation in the genitive absolute introduces an anecdote regarding Plotinus's method of teaching. "When for three days I, Porphyry, had been questioning how the soul is conjoined with the body, he continued demonstrating" (13). Porphyry goes on to relate how a visitor named Thaumasius expresses his impatience with such an extraordinarily digressive method: "He wanted to hear Plotinus make general statements and as in treatises, but he couldn't bear Porphyry responding and questioning." With palpable satisfaction, Porphyry reports that Plotinus replies as follows: "But if we could not solve the *aporias* of Porphyry's questions, we would not be able to say anything at all in the treatise" (13). For all of his apparent desire to package Plotinus's wisdom into well-edited texts destined for broad consumption, Porphyry here distances himself strongly from the interloper Thaumasius, who wants to hear Plotinus deliver a coherent lecture. Later, Porphyry will reminisce: others "considered [Plotinus] a dull babbler and held him in contempt because they could not understand what he was saying. . . . He seemed like one conversing with his companions, and the syllogistical necessity embodied within his reasoning was not immediately apparent to anyone. I, Porphyry, experienced similar things when I first heard him" (18). Now, however, he revels in Plotinus's dialogical meanderings, his willingness to defer closure seemingly infinitely—his seductive elusiveness. Following Plotinus, Porphyry is beginning to learn that "knowledge is like an animal—it can be tracked but not captured," as Miller phrases the Socratic insight.[37]

But as for Me, Who Am I Following?

"Plotinus, the philosopher of our times, was like one ashamed that he was in a body"—so reads the striking opening line of Porphyry's text (1). From the start, Porphyry's Plotinus is involved in a game of hide-and-seek with the reader, flickering at the edges of our vision. He is not ashamed of his body but he is like one who is ashamed. (He is not unlike Derrida, then, caught in the gaze of his cat.) Porphyry is offering us a likeness, yet a physical likeness is precisely what Plotinus refuses when his follower Amelius wants to have his portrait made. Amelius commissions one anyway, on the sly, but we do not get to see it; and Porphyry, who spent six years with his teacher, never tells us whether he was short or tall, fair or dark, fat or thin. His textual likeness is of another sort, as Miller shows, yet for all of that it is scarcely disembodied. One could say that the Plotinus of Porphyry's *Life* eludes the reader because he is always departing from body; yet it may be equally true that he eludes us because he is so very embodied.

The first positive things we learn of Plotinus, in swift succession, are that he was frequently constipated yet refused enemas; that he likewise refused medicines from wild beasts, as he did food from the bodies of domesticated animals; that he abstained from bathing in public, preferring the service of a private masseur; and that he died of acute diphtheria—not a pretty death. In his final days, he lost his clarity of voice and sight, and "ulcers formed on his hands and feet"; even his friends withdrew from his company, perhaps because he still insisted on greeting them with a kiss (2). This opening portrait combines gestures of bodily modesty—Plotinus's refusal to be physically penetrated by enemas, medicines, or even the public gaze—with an unflinching account of the inexorable corruption of the body from within. Plotinus will not consume the flesh of other animals; ultimately his will go the way of all flesh.

The anecdote that Porphyry recounts immediately following his report of Plotinus's illness and death is one that Plotinus himself had shared with his companions, we are told. Quite different from the report that precedes it, it is nonetheless similarly intense in its physicality: "Although already attending grammar school, being as old as his eighth year, he used to visit his nurse and, baring her breasts, he wanted to suck; but being told one day that he was a mischievous child, feeling abashed, he restrained himself" (3). Not surprisingly, Porphyry's decision to include this passage has left most commentators stumped, or even embarrassed, yet its literary effect is quite striking. We have already been told that Plotinus refused to tell "his ancestry, his parentage, or his birthplace" (1). Instead, he volunteers a kind of belated infancy narrative, presenting an image far more arresting than that of a helpless nursling—a school-aged boy who actively desires to suckle. The figure is not only regressively infantile but also, and more important, precociously erotic. The child is *already* feeding from the font of wisdom, even as he is sent to study with a teacher of the elementary arts of grammar—skills that he clearly disdains as an adult, long after his disciples might have expected more mature behavior. As Porphyry recounts, "For when he used to write, he did not form the letters with a care for beauty or divine the syllables clearly or pay attention to spelling but adhered only to the thought and, what we all marveled at, he continued this practice until the end" (8). It is difficult to discipline Plotinus: he is like a wild beast in the text; even the narrative of his life will not obey chronological order.

The only explicit description of Plotinus's physical appearance is the following: "When he was speaking his intellect visibly illuminated his face." Luminosity is a stock characteristic of holy men in late antiquity, but Porphyry's version gets more physical than most. He continues: "Always of winning presence, he became at these times still more engaging: a slight moisture gathered on his forehead; he radiated benignity" (13). What is the radiance that betrays the beauty of Plotinus's inner soul? Porphyry discovers it in the sweat of his brow.

The Philosopher That Therefore I Am

"It is extremely rare to have the chance to see someone in the process of training himself to be a human being," writes Pierre Hadot, with reference to Marcus Aurelius.[38] Here, I have suggested that relations of following and traces of animality in Porphyry's *Life of Plotinus* may, alternatively, allow us to perceive someone in the process of training himself to be *more than,* or even *other than,* a human being.

As Porphyry's opening avowal warns, Plotinus's animal body *appears* to be an embarrassment to the orders insinuated by the very bio-bibliographical text that constructs it as such; it breaks out of both the linearity of narrated time and the stasis of eternal ideas as vivid but elusive presence. In that sense, it does not follow; it is not sequential; it disrupts convention. But perhaps that is just the point. The Plotinian body gives witness that following is not a matter of imitating, philosophy not the production of more of the same. Porphyry can brag that Longinus recognizes that he, more than any other, took Plotinus as his model (21). What does it mean, however, to take as your model one who follows no model, as Porphyry emphasizes (14)? What happens to the autobiographical gesture of the *je suis*, the *ego Porphyrios*, when being is following, and the philosophical animal that one is following slips away like a serpent, exulting to join the divine in himself with the divine in all the animals?

Let me close this reading with a final anecdote from the *Life*. On the occasion of Plato's birthday (celebrated by Plotinus in lieu of his own), Porphyry reads a poem entitled "The Sacred Marriage." It abounds in erotic teachings so enigmatically encrypted that some of his audience say he has gone mad, but Plotinus praises him as having achieved the status of poet, philosopher, and priest, all at once. However, when on another day another speaker suggests that a disciple should, "for the sake of acquiring knowledge of virtue," play Alcibiades to his master's Socrates by "submitting to the teacher as lover, even to the extent of sexual intercourse," Plotinus is deeply disturbed. At his request, Porphyry offers a refutation, and the master is so delighted by his disciple's discourse that he cannot contain himself; he quotes the Homeric line repeatedly, "Keep striking like that, that you may be a light to men" (cf. Iliad 8.282) (15). Thus the eroticism of Plato's symposium is banished by the excitement of Homer's battlefield, with Porphyry wielding Teucer's famous bow, Plotinus cheering him on like Agamemnon. But why is Plotinus so excited? Does the philosopher rejoice to see his disciple reinstate the human as other than animal, as predator rather than prey, therefore also as master of his own carnal desires? I would, of course, prefer to read it otherwise. Perhaps what delights Plotinus (who is after all a vegetarian) is Porphyry's refusal of the distinctly *human* order of predation that reduces flesh to mere meat. Knowledge of virtue cannot be acquired in a vulgar exchange for sexual pleasure. No, the hunt must go on. The mutual seduction of animal bodies remains in play for those who would follow Plotinus.

Queerly Ecological: The *Lives* of Antony, Paul, and Mary of Egypt

The ascetic Christians who follow Plotinus must be tracked across the deserts of Egypt and Palestine. Athanasius of Alexandria, writing in the mid-fourth century, pursues his prey from a distance: most of what he knows of the elusive monk Antony he has learned from others. Tracing his hero's journey into the desolate mountains east of the Nile, he produces the first Christian hagiography; in so doing, he both imitates and competes with his pagan predecessor Porphyry's biographical writings.[39] Roughly two decades later, the Latin writer Jerome seeks in turn to outdo the Greek *Life of Antony* with his own *Life of Paul*, featuring a fictional Egyptian hermit who preceded Antony (or so Jerome claims) and whom Antony himself must track. From this point, literary lives of saints, ever a fluid and eclectic genre, swiftly multiply and spread. In the early seventh century, Sophronius of Jerusalem authors a Greek work that is clearly indebted to Jerome—the *Life of Mary of Egypt*, in which the monk Zosimas plays Antony to Mary's Paul, pursuing her into the depths of the Transjordanian desert.[40]

These three texts will be the basis of my initial sketch of the queer ecology of late ancient hagiography. Several features particularly interest me—the queering of time, the queering of place, the queering of the human, and (running through all of that) the queering of desire.

Queering Time

In a sense, even to read ancient Saints' *Lives* into our own anthropocene moment is to open a kind of queer temporality, creating "a friction of dead bodies upon live ones, obsolete constructions upon emergent ones," as Elizabeth Freeman puts it. Freeman dubs this "historicist *jouissance*" or, better yet, "temporal drag."[41] Saints are drag queens and kings of time, then. They have a way of disrupting normal, and normative, chronologies. Refusing to be confined to their proper historical moment, they never completely leave it either: hence the whiff of anachronism that enhances their exoticism. But there is another sense in which Saints' *Lives* queer time, as I have already hinted. They do so by bending the genres of romance and biography. Ancient novels typically feature a "boy meets girl" plot: initially stricken by love, a young man and woman are subsequently parted but ultimately reunited and married, incorporated into their proper adult roles. However, the Saint's *Life* embodies "the potentiality of a life unscripted by the conventions of family, inheritance, and child-rearing," to borrow Judith Halberstam's words.[42] Biography typically charts the advance of a statesman's or a philosopher's career and character. But famously, the saint fails either to produce or to progress.

When decades roll by in which, apparently, nothing of note occurs, should narrative time be said to pass quickly or slowly? Or does it, rather, fold in on itself so that the reader both steps lightly across the gap (as if time had sped up) and also confronts the void of the fold itself (as if time had dropped out altogether). Intriguingly, this empty, hollowed-out time is often the time of the saint's elided adulthood: we may meet the saint as a child or a youth only to encounter him or her again in old age. Halberstam has explored the "stretched-out adolescences of [contemporary] queer culture makers that disrupt conventional accounts of sub-culture, youth culture, adulthood, and maturity."[43] For the Christian saint it is likely to be a stretched-out senectitude that marks the queerness of time's fullness. Stretched out and yet also all too compressed. For when we meet the saint, time is already running out. As Halberstam puts it, with reference to the AIDS epidemic, "While the threat of no future hovers overhead like a storm cloud, the urgency of being also expands the potential of the moment and . . . squeezes new possibilities out of the time at hand."[44] Similarly, the Saint's *Life* draws us into a moment of condensed possibility—a moment that may harbor the bittersweet pleasure of a long-deferred erotic encounter between a beloved teacher and a desiring follower.

In comparison with the *Life of Plotinus*, the *Life of Antony* seems straightforwardly linear: a narrative strand stretches from birth to death, strung with anecdotes and well-formed discourses. Whereas Porphyry could not supply Plotinus's "ancestry, his parentage, or his birthplace" (Life of Plotinus 1),

Athanasius is able to provide a proper account of beginnings, recording that Antony was an Egyptian born to noble and well-to-do parents. However, the plot quickly veers off course. Like Porphyry, Athanasius describes his hero as an unusual boy, a docile homebody uninterested in school, friends, or even tasty treats (Life of Antony 1).[45] This queer childhood is a hint of things to come. When his parents' death leaves him the head of the household at age eighteen or twenty, Antony does not do what one would expect—marry and produce an heir. Instead, he abdicates both property ownership and the guardianship of his sister (2–3), thereby "challenging conventional logics of development, maturity, adulthood and responsibility," to borrow Halberstam's words again.[46] As Athanasius sums it up, "Henceforth he had leisure [ἐσχόλαζε] for ascesis rather than the household, turning his attention to himself and training himself patiently" (3). Opting out of re/productive time, Antony submerges himself in a time of idleness (σχολή); his only work is his own life.

Or perhaps it would be better to say, as Edith Wyschogrod suggests, that the saint does not *work* at all—a term she associates with Heideggerian notions of human manipulation and mastery, the use of tools and discourse, and the activity of "dwelling." In this eschewal of work, saints are like animals: "Having no history apart from species continuity, they develop simple, unmediated relations to their environment. Without stretching the term too much, they can be said to labor rather than to work." Labor, in contrast to work, sustains life in and through bodies; it sits close to the generativity as well as the vulnerability of flesh; it does not proceed toward a predetermined goal; its temporality is not progressive or linear. Wyschogrod adds, "To accept corporeal vulnerability by divesting oneself of home and history so far as possible is to transcend the essence of man through its underside by taking on sheer animal sentience."[47] Thus, saintly time is animal time. It is not historical; it is not teleological. Rather, it is open time—a time of waiting.

A time of waiting for the arrival of another, perhaps. In Antony's life, the other who arrives is sometimes a teacher, sometimes a disciple, sometimes a wild beast or a demon. Most days no one comes, but near the end he keeps a little garden just in case (50). How does one write the life of one who waits, who lives in open time? Rather than a smooth tale of progress, the *Life of Antony* offers a loosely assembled series of snapshots that suggest not so much a development as an intensification of the virtues already present in his childhood. We see Antony as a young man patiently observing other monks on the outskirts of his village (4–7); living alone first in a tomb (8–11) and then in a deserted fortress into middle age (12–14); interacting with other desert dwellers and visitors (15–48); and finally withdrawing to a mountain in the eastern desert (49–93). The bulk of the *Life* lingers, and very nearly stalls, in the half-century between Antony's 55th year, when he emerges from the fortress in which he has been enclosed for twenty years, and his 105th year, when he dies on the mountain. Athanasius impresses upon his readers how little Antony has changed over these many decades, offering a telling detail: "He lost none of his teeth—they simply had been worn to the gums because of the old man's great age" (93). Like the landscape itself, Antony endures, the contours of his body only gradually reshaped by the gentle but persistent force of his own rumination. Even the cycle of human generations has slowed. Little remains for the saint to bequeath—two sheepskins, a cloak, a hair shirt (91), and a reputation for extraordinary piety (93). The two monks who are the companions of his old age bury his body in a secret place. His biographer Athanasius, inheritor of one of the sheepskins and the cloak, does not know where he lies.

Jerome's *Life of Paul* does not begin with the hero's childhood, as the *Life of Antony* does. We initially encounter Paul at age sixteen, an Egyptian youth well educated in both Greek and Coptic who has just come "into considerable wealth at the death of both of his parents," not unlike the Antony of Athanasius's *Life* (Life of Paul 4).[48] And here Paul's life, like Antony's, takes an unexpected turn, yet the turn is not of Paul's own making. Crisis impinges from without, as greed rears its ugly head. It is a time of persecution, and Paul's brother-in-law determines to betray Paul, a Christian, to the authorities, so that he may seize the family's wealth (4). Perceiving this plot, Paul flees to a deserted location in the mountains to wait out the persecution (5). There he makes his home in a cave, spending his time "in prayers and solitude," receiving both food and clothing from a palm tree (6).

Jerome has already warned us in his prologue that "no one has discovered how [Paul] lived in his middle years" (1). Nonetheless, it is a surprise when, after a brief digression regarding the plausibility of Paul subsisting on nothing but dates, Jerome resumes his narrative at a point when Paul is no less than 113 years old—elderly and then some (7). Ninety-seven years have been swallowed in narrative silence and the presumed tedium of "prayers and solitude." We may wonder at what point Paul decided that there was no going back to the life promised him as a youth: that moment has surely long since been forgotten, even by Paul himself. Paul is now an old, old man, but we will not meet him just yet. Having established Paul's place and his age, Jerome shifts the narrative focus to another old man living in solitude in the desert.

Antony, who is merely ninety, has been beset by a terrible suspicion, namely, "that no more righteous monk than he dwelt in the desert" (7). There is no one, that is, from whom he can learn. Happily, a dream reveals that his fear is unfounded, and Antony sets out immediately the next morning to find the solitary dweller who is his superior—though he has no idea where this holy person lives. Guided by a series of miraculous messengers, he finally arrives at the entryway to Paul's cave, utterly exhausted. After a bit of coy hesitation, Paul allows Antony to enter. "The two embraced, greeted each other by name, and then together gave thanks to God. After bestowing the sacred kiss, Paul sat down beside Antony and began to speak thus, 'Look at the gray hair covering limbs decayed with age. This is the man whom you have sought with such hard work. You see before you a man soon to become dust. But . . . love endures all'" (9–10). The elders gossip happily; they share a meal of bread and water and then spend the night joined in prayer. In the morning Paul says to Antony, "My brother, long ago did I know that you were living in the desert. Long ago did God promise you to me as my fellow servant. But now the time has come to sleep and . . . with the race completed, there remains for me the crown of righteousness. You have been sent by the Lord to cover my wretched body with soil, returning earth to earth" (11). Long ago: *how* long ago? How long *has* Paul been waiting for Antony? And now that Antony has finally arrived, he sends him away after a mere day to fetch a cloak in which to bury him. Antony begs Paul to let him stay and die with him, but Paul is adamant. By the time the grieving Antony has returned, the old man is dead. All that remains is to bury him, as he has requested.

Sophronius's *Life of Mary of Egypt* follows a similarly queer timeline, although the narrative is once again somewhat differently structured. Rather than beginning with the story of Mary, Sophronius focuses initially on her soon-to-be-disciple Zosimas; the account of Mary's early years is presented as a flashback, related by Mary after the two have met. Like Jerome's Antony, Zosimas has reached old age and fears that he is surpassed by no monk on earth (Life of Mary 3).[49] In response, he receives divine instruction to go live in a monastery near the Jordan River, where the monks have the practice of spending Lent in solitude in the desert. When Lent arrives, Zosimas turns eagerly toward the innermost part of the desert, hiking for twenty days in hopes of "finding a holy father dwelling there who could help him discover what he longed for"—namely, to become the disciple of one who outdoes him in ascetic practice (9). On that twentieth day he sees someone—or something—in the distance. Filled with joy, he races toward the creature, the first living being he has seen since he began his desert journey. The creature flees; the old man chases. Finally, they stop on either side of a dry streambed. The creature refuses to turn toward Zosimas but does call him by name, declaring shame at her nakedness—for the creature is both human and female. After Zosimas tosses her a cloak, the woman blesses him and then, like Jerome's Paul, asks him for news of the world. Zosimas complies and in turn asks for her story.

Reluctantly, she relates the narrative of a twelve-year-old runaway girl who lived a life of sexual promiscuity in Alexandria for seventeen years. Her time was always queer time, it seems. Then, having jumped on a ship to Jerusalem for the fun of it (the sailors were handsome), she has a strange and moving experience at the doors of the Church of the Holy Sepulcher. As a result, she becomes a devotee of Mary, the Mother of God, who instructs her to seek her dwelling place in the land beyond the Jordan. Bringing with her only two-and-a-half loaves of bread, Mary takes up residence in the Transjordanian desert. She has been there for forty-seven years. Zosimas is beside himself with curiosity, but Mary can tell him little of her life as an ascetic: it has apparently been uneventful. For the first seventeen years she struggled mightily with her desires for wine and food, song and sex, but the Mother of God gave her strength (28–29). She has fed herself on wild plants "and whatever else can be found in the desert" ever since her bread ran out (30). Her clothing has long since disintegrated. And she has seen no human or other animal in all her years in the desert, until meeting Zosimas (31).

Having told her story, Mary sends Zosimas away and she herself runs deeper into the desert—but not before she has arranged for him to meet her in a year, on the bank of the Jordan on the night of the commemoration of the Last Supper, to offer her holy communion. On the appointed night, Zosimas prepares himself and sits on the bank of the river, waiting. When Mary reaches the opposite bank, to his astonishment she simply walks across the surface of the river. Giving the joyous Zosimas "the kiss of love on his mouth," she receives the sacrament and then asks that he come in yet another year's time to the dry riverbed where they initially met (35). Expressing his wish that he could be with her forever (but seemingly knowing that he cannot), he agrees, and she departs. When Zosimas returns to the riverbed the next year, he finds Mary's corpse; she died on the same night that he gave her communion. All that remains is for him to bury her.

The life of the saint does not, then, run a smooth course from youth to adulthood, much less to marriage and children. Its time is not the time of reproduction or of maturation. Its time is not productive at all. It is the time of old men and women: frequently nothing much happens. Long periods of waiting—of prayers, silence, and solitude—are interrupted by brief intervals of connection and joy. Finally, there is only the "time at hand," the moment of searing bliss; for Paul and Mary, it marks the meeting of teacher and disciple. Then death. And after death, what remains? The disciple remains. The text remains, for as long as there is a reader.

As Morton observes, "Environmentalism is often apocalyptic. It warns of, and wards off, the end of the world." He adds, "But things aren't like that: the end of the world has already happened. . . . Today is not the end of history. We're living at the beginning of history. The ecological thought thinks forward."[50] At the same time, and paradoxically, he declares that "there is no future."[51] That is to say, there is no more deferral, no more living—or sacrificing—"for" the future; nor is there any clear vision of what lies ahead. Queer theorists have similarly argued for "a politics without a future," suggesting that "to queer things is to transform them in ways that we cannot anticipate."[52] Thinking forward is, then, leaning into unanticipated possibilities while at the same time finding new "ways of thinking about the here and now," as Nicole Seymour urges, adding that "the belief that environmental devastation is a *possibility*, rather than a current and impending reality, or that we have to clean up the planet for *future* generations, rather than for present ones, allows for the kind of complacency that authorizes such degradation in the first place."[53] Thinking forward also involves rethinking what counts as *generations*. As Catriona Sandilands puts the question, "What, given non-reproductive forms of generativity, constitutes a child, and where, in this conception, is there space for Others who are not our children, and perhaps not anyone's children in the way we conventionally imagine progeny?"[54] Finally, thinking forward may sometimes involve thinking backward—or, as Freeman has it, "mining the present for signs of undetonated energy from past revolutions."[55]

Queering Space

Saints' *Lives* queer space as well as time. They do so in part by bending conventional representations of "nature." As literary historian Ernst Curtius famously argues, "the principal motif of all nature description" from antiquity to well beyond is the *locus amoenus*, or "happy place," a "beautiful, shaded natural site," composed of "a tree (or several trees), a meadow, and a spring or brook."[56] The persistent popularity of this literary trope notwithstanding, Mark Edwards has noted that some early Christian writers reject the *locus amoenus* in favor of what is classically framed as a *locus horridus*, or terrible place—namely, the sea. Edwards suggests that for Christians, the sea no longer serves as a figure for the painful vicissitudes of bodily existence or the weary buffeting of fate, but instead represents a primordial wilderness in which God may be discovered, as well as a site of baptismal rebirth.[57] At the same time, however, other ancient Christian texts—hagiographies in particular—make a slightly different move, not replacing the pleasant place with a conventionally horrible one but perverting it, so to speak. It is still a happy place, even a paradise of sorts, marked by tree, meadow, spring, or brook. No longer merely a spot to dally on a leisurely afternoon but now a place to call home, its positive affectivity may seem to have intensified, if anything. Yet the saint's happy place is also shadowed with ambivalence: a horror, a haunting, a sense of the uncanny has intruded upon the imaginary of the pastoral landscape.

We first encounter the hagiographical happy place in the *Life of Antony*, where Athanasius describes a site located in what he refers to as "the innermost desert" (49.4). "Journeying three days and three nights with [his Saracen guides], [Antony] came to a very high mountain. And under the mountain there was water most clear, sweet and very cold. Outside there was a plain, and a few untended date palms. Then Antony, as if divinely moved, fell in love with the place. . . . He remained alone in the mountain, no other living with him. Recognizing it as his own home, henceforth he stayed in that place" (49.7–50.2). This scene includes the "minimum ingredients" of the *locus amoenus*, as Curtius lists them—"a tree (or several trees), a meadow, and a spring or brook." Curtius notes, "Birdsong and flowers may be added. The most elaborate examples also add a breeze."[58] But Antony's happy place seems to subtract rather than add. Its "several trees" are "few" and "untended" palms (49.7), scraggly specimens—although modestly productive, affording him "some meager and paltry comfort" (50.4) as supplement to a diet consisting otherwise only of bread. Its "meadow" is a desert plain (49.7), otherwise bereft of vegetation—although Antony does subsequently manage to find just enough arable ground to grow grain for his bread, as well as a few vegetables for his visitors (50.6–7). Its "spring," welling forth at the foot of the rocky mountain that shelters Antony (49.7), attracts wild animals who trample his crops—leading him to order them to leave the oasis (50.8–9). In short, this is a place of fragile sufficiency rather than lavish abundance; it provides not a backdrop for civilized leisure but hard-won sustenance for bare life.

Antony's happy place not only subtracts but also adds, however. It adds not flowers but stalks of grain and some vegetables. It adds not birdsong but the unnerving howl of hyenas and the infernal racket of demons. His visitors "heard tumults and many voices, and crashing noises like the sound of weapons; and at night they saw the mountain filled with beasts. . . . And it was truly amazing that being alone in such a desert [Antony] was neither distracted by the demons who confronted him, nor was he frightened of their ferocity when so many four-legged beasts and reptiles were there" (51.3–5). The line between beasts and demons blurs, as do the contours of Antony's solitude. The innermost part of the uninhabited desert enfolds multitudes, it seems. Thus the austerity of subtracted sufficiency is matched by a certain creepy-crawly extravagance.

Here we have what Alfred Siewers refers to as a "literary geography of desire" that complicates the "delight in nature" characteristic of previous literary tradition. The mountainous terrain is not merely *pleasant*; instead, it inspires both trepidation and an intense sense of connection. As Siewers puts it (with reference to early medieval Irish literature), the saint is struck by "terror at the chaos of nature mixed with awe at the transcendent divine, alongside experience of both an immanence of the divine and an intimate sense of place in landscape."[59]

In her essay "Happy Objects," Sara Ahmed explores "how happiness functions as a promise that directs us toward certain objects, which then circulate as social goods. Such objects accumulate positive affective value as they are passed around. . . . Affect is what sticks, or what sustains or preserves the connection between ideas, values and objects."[60] In particular, Ahmed is concerned with "how family sustains its place as a 'happy object,'" and she suggests that this is accomplished in part "by identifying those who do not reproduce its line as the cause of unhappiness." She dubs these others "affect aliens," and they include in this case "feminist kill-joys, unhappy queers, and melancholic migrants."[61] From Ahmed's deliberately donned "kill-joy" perspective, the promise of happiness must sometimes be refused.

Places—not least imaginary places—are among those "objects" to which affect can stick. As environmental critic Lawrence Buell notes, "One also becomes attached to places by the power of imaging alone. . . . The places that haunt one's dreams and to some extent define one's character can range from versions of actual places to the utterly fictitious."[62] The literary topos of the *locus amoenus* is a prime instance of such an affectively sticky imaginary place: pleasure, delight, and happiness cling to its shady groves and babbling brooks. Circulating as the signifier of a social good, the *locus amoenus* underwrites elite privilege, aestheticizing and commodifying nature while also naturalizing desire—perennial ploys of literary pastoral. The *Life of Antony*, I want to suggest, implicitly refuses the particular happiness promised by the *locus amoenus*, much as it explicitly refuses the happiness of family life (a not altogether unrelated "social good"). However, it does so less by foregrounding the position of the "unhappy queer," for example, than by queering both place and the happiness that sticks to it. This queering of place and affect brings us to the brink of ecological thought, not only because it may offer what Scott Hess refers to as a "sustainable pastoral"[63] but also because it introduces the uncanny into the imagination of place.

When Antony reaches the inner desert, a terrain with which he has already pronounced himself inexperienced (ἄπειρος) (49.5), he is struck by its unexpected familiarity. "This was [the place] that the one speaking to him on the bank of the river had indicated" (50.1), he realizes immediately. More than that, "he recognized it as his own home," we are told (50.2). Here Morton's suggestion that "the essence of the local isn't familiarity but the uncanny, the strangely familiar and familiarly strange" seems especially apt.[64] Nature, like literature, loves to repeat: indeed, a literary *topos*—or *commonplace*—is the effect of repetition, of the reworking of already familiar material. In literature and life, place is haunted by its own intertextuality—by the ghostly presence of other places, other trees, meadows, springs, and brooks. "*Here* is shot through with *there*," as Morton puts it.[65] To be sure, the classical *locus amoenus* strives for containment nonetheless; as Petra Hass emphasizes, it is characteristically limited to a "coherent, manageable portion of the landscape" that includes some kind of shelter.[66] By contrast, Antony's queer happy place—his *unheimliches Heim*—seems to embrace the uncontainability of place. We may *imagine* that he lives in an enclosed cave, but the text only mentions the mountain, the water, the plain, the palms, a patch of dirt cultivated with simple tools. Inside the outside, everything opens up.

Antony's uncanny home is haunted by more than the ghostly presence of other places, by more than a sense of the way places open onto one another. It is haunted too by an awareness of what, as we have seen, Morton calls the "mesh"—the intimate entanglement of creatures with one another, the irreducibility of interconnectivity. *Eremos* indicates an uninhabited region—a place that is desolate or lonely. It is a place defined by a kind of *dis*placement or *spacing*, then—by an emptying out. But it also *makes space* for movement and relationship, as Antony comes and goes, and so do his visitors. Crucially, encounters with other others are also staged there. Here again, Morton's comments resonate: "Even if we are alone in the 'wilderness,' we are not alone."[67] Having evaded the troubling crowding of *human* community, Antony comes face to face with *something*, but it is hard to say what. He is said simultaneously to be alone (μόνος) and to be surrounded by demons and wild beasts (51.5). His visitors hear noises, see visions, but only Antony engages them. Are they animals or spirits? Enemies or allies? Other or self? Morton continues with his description of the uncanny experience of deserted places as follows: "What is scary about being lost in a forest of tree upon tree . . . is catching a glimpse of yourself, from the point of view of the trees. It is the feeling of being watched, of being accompanied. . . . The dark openness gazes at us. We aren't exactly seeing ourselves in a mirror. We're seeing ourselves as the void that looks back at us."[68] Ancient representations of forests as dreadful places resonate with this description: as Aneta Kliszcz and Joanna Komorowska note, there one finds oneself "in a world inhabited by powers far beyond human comprehension . . . , beyond human experience, untamed, and, even worse, untamable."[69] According to the *Life of Antony*, one night "nearly all the hyenas in that desert, emerging from their dens, surrounded [Antony], and he was trapped in their midst" (52.2). They threaten to bite; surprisingly, they flee without biting. On the inside of the outside—*khora*, the place of Antony's *anakhoresis*—things are not always what they seem. Or rather, the borders between things do not quite hold. A saint may see himself as the hyenas that look back at him.

As Curtius notes, the *locus amoenus* traditionally "formed part of the scenery of pastoral poetry and thus of erotic poetry"; he cites verses of Petronius in which a shaded grove with babbling brook, violets, and birdsong is described as "*dignus amore locus*" (Carmen 131).[70] Jerome's *Life of Paul* opens with a lavish, classicizing description of just such a "place fit for love," which becomes the site for a very strange form of torture. "The devil ordered another Christian, one in the flower of manhood, to be led away to the most delightful [*amoenissimos*] gardens. There, among the radiant lilies and blushing roses, next to a gently murmuring brook, while the wind softly whispered among the leaves of the trees, the youth was placed upon a bed of feathers and, so that he might not escape, bound with caressing garlands and then left alone" (Life of Paul 3). This scene includes all of Curtius's "minimum ingredients" of trees, meadow, and brook. It also includes Curtius's "additional" elements of flowers and breeze. This sets the stage for a bit of B&D sexual theater, as Jerome goes on to describe how a beautiful prostitute interrupts the bound youth's solitude, embracing his neck and fondling his genitals. "And when his body was aroused with passion, the wanton conqueror threw herself on top of him." The boy's response? "He bit off his own tongue and spat it in the face of the one kissing him" (4). Pain allows him to prevail over the seductive pleasures of the conventional happy place, now refigured as a scene of thwarted seduction and refused temptation.[71] The tongue-biting youth would seem to be an "affect alien" or killjoy if ever there was one.

Yet this is not the end of the story. The account of the youth in the garden is presented as a prelude to the tale of the youthful Paul, who finds his own happy place while fleeing the threat of persecution, as we have already noted. Let us now take a closer look at the scene of discovery. "Finally [Paul] came upon a rocky mountain, not far from the foot of which was a great cave closed off by a stone. . . . He removed the stone and eagerly explored the interior, where he came upon a large room open to the sky. Here an ancient date palm formed with its spreading branches a ceiling. At the base of the palm was a crystal-clear spring, from which flowed a brook. The earth that gave birth to the brook soon swallowed it up through a small opening. . . . Paul fell in love with his dwelling, as if it were given to him by God" (5–6). Jerome offers us a more visually detailed and coherent description of the hermit's new home than does Athanasius; he also describes a place more contained than Athanasius's scene. The cave is clearly depicted as an enclosure, its entryway sealed by a rock, its upper opening sheltered by a palm tree that provides Paul "continually" with both food and clothing (6). Thus the tree, spring, and brook of the classical *locus amoenus* are all encompassed within the cave, in a queer inversion. At the same time, the enclosed intimacy of the mountain cavern is countered by the open expanse of the rocky terrain that extends beyond it, empty of human inhabitants—but not empty of life altogether, as Antony discovers when he sets out to find Paul. The "earth" (*terra*) is said to "give birth" to a brook within the cave, while outside it, the "wilderness" (*eremus*) typically produces "monstrous animals"; it is capable of "bringing forth a beast" (7). As Antony travels toward Paul, his gaze takes in both "the tracks of wild animals and the vast expanse of the desert" (9).

Jerome's description of Paul's discovery of his own desert dwelling thus mimics and revises both classical and Athanasian versions of the happy place (5–6). The echoes of Antony's mountain home may be the more obvious, particularly when one compares the description with Jerome's rendering of

Antony's dwelling in his *Life of Hilarion*: "There is a high and rocky mountain extending for about a mile, with gushing springs at its foot, the waters of which are partly absorbed by the sand, partly flow towards the plain, and gradually form a brook shaded on either side by countless date palms which contribute much pleasantness (*amoenitas*) and charm to the place" (Life of Hilarion 31). Here, as in his description of Paul's cave, Jerome has made even the ascetic *locus amoenus* lush by comparison with Athanasius's more austere *topos*. However, Paul's cave does not only recall Antony's; it is also haunted by the luxuriant gardenscape of the *Life*'s initial tale, not only because of the intimacy of its enclosure and the fertility of its plant life, but also because of the erotic pleasures it will enfold in the long-awaited encounter between Paul and Antony, as we have seen.[72]

When Antony leaves and returns to find that Paul has died, the inside opens onto the outside, cave onto desert, as enclosure becomes exposure. Wrapping Paul in the cloak he has brought, Antony carries the body outside the cave, singing psalms and hymns. Paul has told Antony that he is "a man soon to become dust," and now Antony covers his dust with dirt, as Paul has directed him to do. At the same time, Antony clothes himself in Paul's own rough tunic, woven by the saint from palm leaves (Life of Paul 16). Earth to earth: Paul has become part of the land. Antony returns to his own desert cell, wrapped in a tactile memory of Paul's palm-shaded happy place.

The *Life of Mary of Egypt* opens the space up further. Its wilderness offers no shelter for its saintly subject, whose location is no more specific than "beyond the Jordan" (Life of Mary 25–26) or "the innermost part of the desert" (11): this holy woman is always on the move. The first and last encounter of Mary and Zosimas is in a wadi, or dry riverbed, as we have seen (11, 37). There is no spring, no palm—no protective enclosure, no dates for food, no leaves for weaving tunics. And there Zosimas, like Antony, must bury his beloved. "Return 'dust to dust,'" Mary requests in a mysterious message written posthumously in the dry, dry dirt (38). Obediently, Zosimas covers her nearly naked body with earth, returning to his monastery to tell her tale.

Here the saint's *Life* queers the *locus amoenus,* or happy place, almost beyond recognition. Its pleasures are austere, its beauty stark and uncanny, its expanse unbounded. Yet can literary pastoral ever be *queer enough*, from an ecocritical perspective?[73] As Greg Garrard puts it, in a seminal 1996 article, pastoral "may cloud our social vision, or open out a human ecological one; it may help in the marginalization of nature into 'pretty ghettoes' . . . or engender a genuine counter-hegemonic ideology."[74] More recently, Laura Walls has called for a radicalizing of pastoral that will "insist that the local is already and has always been planetary, just as the global has always been under our feet." She adds that "there is something radically strange and wild . . . that is still very much a part of the pastoral ideology."[75] Anthony Lioi describes the promise of "an integration of the pastoral" that "leads to a moment of cosmic connection."[76] For Paul Outka, "pastoral space," which is at once "artificial and natural," might "provide the ground of a post-Natural environmental politics going forward. . . . We should see it as . . . a place of instability and mixing that has to be produced in an ongoing way," he urges.[77]

In their transformations of the classical "happy place," our three Saints' *Lives* seem to offer such spaces of radical wildness, perched at (or beyond) the furthermost border of habitable land; their desert landscapes do not close in on themselves but open up to the very cosmos; and distinctions between urban and rural and artificial and natural break down. Athanasius famously declares that Antony's desert became a city (Life of Antony 14), while Jerome notes that Paul's cave included "several small rooms wherein one could see mallets and long-since-rusted anvils," remnants of a secret mint from the time of Cleopatra and Antony (Life of Paul 5). These are places neither pristine nor static. They are places with their own histories—haunted, even. They are places constituted by the lives that traverse them. They are places of ongoing and unpredictable encounter, in which nothing is natural, or else everything is, and every locale is lovable, or else no locale is. As Whitney Bauman proposes, "Rather than an environmental ethic of place, which smuggles in some heterosexist assumptions about the importance of fidelity to a single place and ideas about pure nature toward which we can return, we will need a bit of polyamory toward places. We need, in other words, to love multiple places in order to understand their planetary connections."[78] We need to love like nomads.

Queering the Human

Saints' *Lives* do not only queer time and space; they also queer the human. Indeed, this might well be the defining characteristic of hagiography—its interest in transgressing the limits of humanity. This is achieved in part through a movement of intensification. Extravagant figures and also figures of extravagance, saints are literary monstrosities—ominous portents, oversaturated signs, abysses of meaning. They are human and then some, we might say; by exceeding normal limits, they may no longer seem to be human at all. As Patricia Cox Miller discusses, saints are frequently compared to both angels and nonhuman animals, suggesting less that spirit is opposed to body than that flesh itself is the site of ascetic transformability: as she puts it, the saint "existed with the animals in the world like water in water, in an intimacy so profound that the animal and the angel were one."[79] Saints exist in profound intimacy not only with (and as) the animal and the angel but also with (and as) the demon. David Brakke discusses "the close, indeed inescapable, relationship between monk and demon": the demon was not only the ascetic's necessary sparring partner but also his or her alter ego, a self necessarily ever to be renounced.[80] This intimacy extends to other beings as well—even to the landscape itself, as we have seen. Antony insists that his body not be embalmed but buried in the earth, and both Paul and Mary likewise request burial, returning "dirt to dirt." The saint is constantly in the process of imaginative self-transformation—of becoming-other—whether human, angel, or demon, animal, vegetable, or mineral.

As Jean-Christophe Bailly writes, "Ever since the dawn of time we have been visited, invaded, traversed by animals or their phantoms."[81] Demons and wild animals figure particularly strongly in the *Life of Antony*, and it is often difficult to discern when Antony is encountering actual animals and when he is encountering demons who take the forms of animals in order to frighten him. Even when the text is explicit in identifying apparent wild beasts as demonic apparitions, the apparitions still seem strangely "real." For example, we read that during Antony's time in the tomb, "the place was filled with the appearances of lions, bears, leopards, bulls, and serpents, asps, scorpions, and wolves"; moreover, each one moves and makes noises appropriate to its kind; finally, they batter Antony's body, leaving him in great pain (Life of Antony 9). If it walks like a duck, swims like a duck, and quacks like a duck, who is to say it is not a duck? The border between "real" animals and demonic appearances is thin indeed, and the same is true with respect to the border between "real" humans and demonic appearances. The devil is able "to assume the form of a woman and imitate her very gesture" (5) and to speak "with a human voice" (6). Demons may even "speak like the devout" (25), "awaken you for prayer, or counsel you to eat nothing at all" (25). They may "pretend to prophesy" (31). They may claim to be angels (35). Again, how is one to know for sure what is "real" and what is not? On one occasion, Antony encounters a hybrid creature, part human, part donkey—an onocentaur. Surely this at least is clearly a demonic apparition, a modern reader might surmise. But no: the beast travels in the company of demons yet does not seem to be a demon itself, at least in Antony's view (53). Despite all the care given to distinguishing demons from others, it begins to seem that for the *Life of Antony* demons are not so much a separate and discrete category of nonhuman creature as a cipher for the radical transformability and ambiguity of human creatures. On the one hand, the instability of the category of the human is threatening: the monk must not give way to mutability but hold fast to his created nature, Antony urges (20). On the other hand, human hope for salvation rests on the possibility of the transformation effected through both ascetic training and divine grace.

The *Life of Paul* performs a particularly vivid and explicit deconstruction of the human-nonhuman binary. As we have already noted, when Antony sets out to find Paul in the trackless desert, he is aided by a series of strange guides. Let us now take a closer look:

> All at once he beholds a man mixed with a horse, called by the poets hippocentaur. At the sight of this he protects himself by marking the sign of salvation on his forehead, and then he exclaims, "Hello! Where in these parts is a servant of God living?" The centaur gnashes his teeth and tries to speak clearly, but only grinds out from a mouth shaking with bristles some kind of barbarous sounds rather than lucid speech. Finally he finds a friendly mode of communication, and extending his right hand points out the way desired. Then with swift flight he crosses the spreading plain and vanishes from the sight of his wondering companion. (Life of Paul 7)

This desert hallucination of a communicative, yet speechless, monstrosity of a man-horse might seem more than enough for any species-blurring purposes; and unlike the onocentaur in the *Life of Antony*, this beast is not menacing but helpful.[82] And yet there is more. No sooner has Antony resumed his journey than he encounters a dwarf, a homunculus, "whose nostrils were joined together, with horns growing out of his forehead, and with the legs and feet of a goat" (Life of Paul 8). Despite his fear, Antony finds himself drawn a step closer to the uncanny creature, who offers him the gift of some dates and identifies himself as a "mortal being"—that is, like Antony, an animal—who has been sent as an ambassador for his tribe (*grex*);[83] his kind are known by many names, as he himself notes—satyrs, fauns, incubi. He is also a fellow follower of Christ, the creature explains, and he offers prayers for Antony, causing him to weep with joy, "marveling all the while that he could comprehend the dwarf's speech" (8). If Antony and the homunculus speak with the same tongue, perhaps they belong to the same tribe. This is a nearly unthinkable thought, and Jerome again interrupts the narrative line: the satyr is gone in a flash, disappearing from sight as quickly as the centaur.

A third figure appears, "a she-wolf (*lupa*), panting with thirst," who crawls toward the foot of a mountain, where she enters a cave. Antony follows her, advancing "step by step" in the darkness, "sometimes standing still." Hearing him, shy Paul shuts and bolts his door. Antony prays for hours on end for entrance, pronouncing himself "known" by Paul, acknowledging his unworthiness, and threatening nonetheless not to leave until he has seen his beloved. "You who receive wild beasts (*bestias*), why do you turn down a man (*hominem*)?" he cries, and the distinction between man and beast, already doubly disrupted by centaur and satyr, dissolves further, even as Antony attempts to reassert his difference—now also inscribed as a *sexual* difference, for it is presumably the she-wolf Paul has admitted (9).

As we have seen, Paul eventually opens his cave to Antony, and the two embrace. Then, suddenly gossipy, Paul demands news with the following words: "Tell me, how is the human race (*humanum genus*)?" (10). The nearly comical question calls attention to a distinction that has become quite unstable—*genus, race*, the *race* that is *human*. What is the *human race* to one whose body is covered in hair, whose flesh is soon to become dust, who lives in a desert populated with monstrous hybrids? What is the *human race* to a couple of queer old saints sharing a simple meal—a loaf delivered by a bird, in fact—in the brief interval of time that remains before the one will die and the other will bury him?

As for the burial, that too enfolds a surprise: for when the time comes, Antony realizes that he does not have the necessary tool for digging a grave. He has labored for love with his feet, wandering aimlessly in the desert until he finds Paul, but he is not much good at working with his hands, it seems—thus, scarcely human, from at least a certain (Heideggerian) perspective. Fresh grief at this lack gives way to fear and wonder. "From out of the deep desert came running two lions with their manes streaming back from their shoulders. . . . Then they began to scratch at the dirt with their feet," digging Paul's grave without shovels or hands (16). Sensing that it is what they desire, Antony offers them his blessing (18).

Sophronius's *Life of Mary* borrows a trick or two from Jerome's hagiography here as elsewhere. But whereas Jerome has Paul proclaim and display his own wildness ("look at the gray hair covering a body decaying with age"), Sophronius shows us Mary's animality through Zosimas's startled eyes. "While he was chanting psalms and looking up to heaven with an alert eye, he saw the shadowy illusion of a human body appear to the right of where he was standing. . . . At first he was alarmed, suspecting that he was seeing a demonic phantom. . . . But . . . he looked again and saw that in fact someone was walking in a southward direction. What he saw was a naked figure whose body was black, as if tanned by the scorching of the sun. It had on its head hair white as wool, and even this was sparse as it did not reach below the neck of its body" (Life of Mary 10). As Patricia Cox Miller points out, Mary emerges here as a hybrid figure.[84] At once human and demonic, she brings together two rather dramatically different (and differently gendered) biblical figures, the dark-skinned bride of Song of Songs 1:6 ("Do not gaze at me because I am dark, because the sun has gazed upon me") and the "one like a son of man" whose "head and . . . hair were white as white wool" from Revelation 1:13–14. At the same time, her nakedness is emphasized: she is without shelter, without clothing, and without any cultivated food source—not even a date palm. She wanders and grazes like a wild beast. Utterly solitary, she sees no human or other animal during her forty-seven years in the desert, until Zosimas tracks and captures her. Indeed, it seems that Mary is holy precisely insofar as she is no longer distinctly human. When she dies, Zosimas like Antony lacks a proper tool with which to make a grave: he wields a small piece of wood in a vain attempt to dig the hard, dry earth. As with the burial of Paul, help comes from a lion—in this case a single lion, as solitary and as wild as Mary.

"Queer ecology may abandon the disastrous term *animal* and adopt something like *strange stranger*," proposes Morton. "How can we ever distinguish properly between humans and nonhumans?" he queries (and is, of course, scarcely the first to do so). Morton adds, "Desire is inescapable in an ecology that values intimacy with strangers over holistic belonging."[85] In the *Lives* of Antony, Paul, and Mary, the saints journey away from holistic belonging and toward intimacy with strangers; away from divisions of species, race, or gender and toward infinite differentiation; away from human society (even monastic versions thereof) and toward elusive, withdrawing, and intensely desirable creatures soon to become dirt.

For all the enormous destructiveness of Christian doctrines of human dominion over nature, Saints' *Lives* suggest that Christianity also harbors another legacy. A sense of the moment as both haunted and open; a sense of place as both intimate and uncontained; a sense of creatures as monstrous, mutable, mysterious, and endlessly seductive; a sense of life as encounter—might this all add up to undetonated energy from a past revolution?

Interlude: Desertification

A landscape is not a static backdrop, though that is how we are often taught to view it. A landscape is "a part of us, just as we are a part of it," as Tim Ingold notes, adding that "what goes for its human component goes for other components as well. . . . In a landscape, each component enfolds within its essence the totality of its relations with each and every other."[86] A landscape is thus particular, relational, and ever emerging. The rate of its collective becoming may be so slow from a human perspective as to be barely perceptible; more rarely, it may change dramatically within a single human lifetime. In either case, the human does not stand outside such processes but is "part and parcel of the world's transforming itself."[87] Indeed, landscapes may be conceptualized as "rogue agents, actively impressing themselves into the embodied histories of organisms."[88] A landscape belongs to time; it has a history; and the humans who traverse it share and are crucially shaped by that history.

A literary work may invite us to enter into a particular landscape, enfolding ourselves in its relations and inhabiting its history, even if only provisionally. Doing so requires an exercise of imagination, sensory, kinetic, emotional, intellectual, and more; it may also require a certain boldness. To enter into an artistically rendered landscape is to open oneself to change. "To live a landscape: one is no longer in front of it, but in it, one passes into the landscape," writes François Zourabichvili.[89] To enter an artistically rendered landscape is also to risk changing the landscape itself: no book remains the same after being read, no painting after being viewed. Even a bored or indifferent regard will have an effect, shutting the work down, slowing its circulation; correspondingly, an animated regard may enliven the work and increase its powers of attraction.

* * *

As I step into Antony's desertscape, my eye travels from the spring at the foot of the mountain to the "few untended date palms" (φοίνικες ἀμεληθέντες ὀλίγοι) starkly visible on the otherwise barren plain that stretches before me

(Life of Antony 49). How many is a "few"? Let us say five. "Untended" suggests that these trees *might* have been tended: they cannot be categorized easily as either "domestic" or "wild."

Humans have cultivated date palms for at least five thousand years; indeed, the histories of date palms and humans have been entwined for so long that no date palm is completely innocent of human contact. "While feral populations, made up of self-sown or abandoned date palms, are common throughout the areas where the species is cultivated today, authentically wild populations, growing in primary habitats, have not yet been positively identified."[90] The origins of the date palm seem to lie in Mesopotamia, but humans have carried them to new regions: the trees now grow across Southwestern Asia, North Africa, and elsewhere. They even thrive in parts of North America. In the second decade of the twentieth century, Paul and Wilson Popenoe traveled to Iraq and endured significant hardship and adventure to bring nine thousand young date palms back to California's Coachella Valley; there ensued a veritable date craze, including the inauguration of an annual International Festival of the Dates during which the California valley was transformed into an orientalizing fantasy replete with scantily clad women, elephants, and camels.[91] Humans have even helped the palms time-travel: astonishingly, a seed preserved for two thousand years at Masada, Israel, was germinated with human aid in 2005; the tree, named Methuselah, has now successfully fertilized a female palm.[92] Humans have played a major role in the reproductive lives of date palms more generally, assisting with pollination and enabling plants to grow from select cuttings rather than seeds, resulting in heartier and more fruitful trees, a numerical dominance of females, and a greater diversity of kinds. In turn, date palms—which require little water and are tolerant of heat—have given humans an astonishing variety of gifts, enabling some to survive in regions that would otherwise be ill suited to human habitation. Their nutritious fruit remains a staple in the diet of many, especially the poor, in the regions where they are grown and also has a wide range of medicinal applications; the seeds can be roasted to make a coffee-like drink or ground for animal feed; the oil of the seeds can be eaten or used on skin and hair; the trunk and leaf stalks, as well as the leaves themselves, can be made into wood, rope, and woven products.[93] Every part of the plant has uses for humans. New ones are still being discovered.

Antony's palms are not cultivated—though they might have been. Like him, they are a bit wild, standing aloof from other humans. Wind carries pollen from the male trees to the female ones, and any new palms will grow from seeds that have landed in the desert soil. The yield of fruit will likely be meager, but Antony does not need much. He is surely too old to climb the tall, branchless trees; thus, he will have to wait for the fruits to drop and hope that he finds them before other creatures do. Perhaps he simply enjoys the shade and humble beauty of the trees; after all, he has bread for sustenance.

Antony's palms evoke other palms, opening his landscape onto other landscapes. We recall that Paul has a date palm in his own cave, also in the eastern desert of Egypt; he eats its fruit and weaves his clothing from its leaves. The part of the Transjordanian desert that Mary roams seems, in contrast, bereft even of palm trees.

I myself tasted fresh dates for the first time just a few years ago, having bought the smooth yellow fruits at a supermarket in Jerusalem without knowing what they were. Nor did I know then that they were most likely grown on the plains that stretch out at the foot of the Judaean Mountains east of Jerusalem, extending to the Jordan River and the Dead Sea. There the date palms are not few but many, and they are not untended but carefully cultivated. The irrigated groves are extensive, the sturdy trees arrayed in rows like soldiers. These palms are firmly embedded in the seemingly unstoppable workings of agribusiness—or what Timothy Morton identifies, more broadly, as the "twelve-thousand-year machination *agrilogistics*."[94] Their fruits, of the coveted Medjool variety, are as politically controversial as they are profitable, for these groves are in the Palestinian Territories but are owned by Israeli settlers. Israel supplies about 75 percent of the world's market of Medjool dates, and according to the "Boycott, Divestment, Sanctions" (BDS) Movement website, roughly 65 percent of that supply comes from illegal settlement plantation and packaging facilities in the West Bank.[95] Palestinian date growers, by contrast, struggle not only with threats to their land but also with restricted access to water and shipping.[96] As water supplies in the region dwindle and the political situation shows no sign of easing, the plight of the Palestinian growers will only become more challenging. Yet still dates remain one of the most feasible crops for water-poor lands, and the demand is high, especially among Muslim populations, where they are valued for religious reasons (Mohammed is said to have broken his fast with dates) and because of their nutritional and medical value. A Palestinian taxi driver once explained to me the various ills—including cancer and heart disease—that could be treated with a diet of dates, so long as an odd number were consumed.

Such striking and versatile trees! Having entered Antony's landscape, I find his palms haunting my own.

* * *

Paul's palm tree provides a central orientation for his landscape, but it is the shadowy spaces extending beyond the sunlit center of his cave that draw my attention as I enter this space. "Inside the hollow mountain there were several small rooms in which one could see mallets and long-since rusted anvils of the sort used to mint money," writes Jerome. Lending his description the aura of historical facticity, he adds, "Egyptian records report that this place was a clandestine mint dating from the time Antony was joined to Cleopatra" (Life of Paul 5). Why has Jerome introduced these particular objects into the landscape of the hermit's dwelling place?

If Antony's date palms straddle the boundary between the wild and the domestic, these rusty artifacts straddle the boundary between nature and culture, living and nonliving, present and past. Mallets and anvils are forged from iron, which is among the strongest and the most common of metals. "Deposits of iron are found almost everywhere," notes Pliny the Elder; one can recognize them by the reddish color of the earth, he adds (Natural History 34.41). And it is to reddish earth that forged iron is ultimately reconverted: the propensity to rust limits this metal's otherwise invincible power, rendering it, like other living things, merely mortal (34.40). The process of corruption proceeds much more slowly than that of human flesh: after four centuries or so, the mallets and anvils apparently still sustain their recognizable forms, even as they begin to bleed into the mountain rock. Nonetheless, Jerome encourages a comparison with human skin, indicating the tools' state of corrosion not as *robiginosus* (cf. Pliny's use of *robigo*, or "rust") but as *scaber*—"rough, scurfy, scabrous," according to Lewis and Short's *Latin Dictionary*.[97] Despite their persistence, it has been a very long time since these were fully vibrant, functioning tools, their hard surfaces pressing softer metals—gold, silver, bronze—into alignment with human values and desires.

Pliny hints that iron bears a moral burden through its propensity to be fashioned into weapons, adding that "a number of attempts have been made to enable iron to be innocent" by turning swords into plowshares, so to speak (Natural History 34.39). The making of coins might not seem an innocent endeavor, however. Athanasius's Antony appears to take a dim view of the monetary economy: when gold is strewn on his path, he rushes by without stopping, thereby demonstrating that "he truly has no regard for money" (Life of Antony 12). We recall, moreover, that Paul's exile to the desert is the initial result of his brother-in-law's avarice, a fact that causes Jerome to reflect critically on the lengths to which humans will go to satisfy their "awesome hunger for gold" (Life of Paul 4). But Jerome also hints that the iron tools have been directed toward even more devious ends than merely minting coins—namely, minting false coins. This occurs in the time of Antony and Cleopatra. Why that time and those people? Pliny, an author familiar to Jerome, notes that Mark Antony "mixed iron into the silver denarius," an act that he juxtaposes with mixing base metals into a counterfeit coin (Natural History 1.46.132). Pliny may refer, with some confusion, to the silver-copper alloys that Mark Antony used in the minting of denarii to pay his legions or alternately to the silver-plated iron forgeries that circulated widely: the line between debased imperial currency and unauthorized minting is blurry. In any case, Jerome here seems to insinuate a comparison between the famous Roman triumvir and his Egyptian consort, on the one hand, and the Roman-Egyptian monastic couple, on the other: Antony and Paul (not Antony and Cleopatra) are the real thing, true rather than false, pure rather than degraded. We note that a silver tetradrachm minted under Mark Antony's authority in 37/33 BCE, possibly in Antioch, depicts Cleopatra on the obverse, Mark Antony on the reverse; strikingly, Cleopatra's features mimic those of Mark Antony's, classically Roman, vigorous and virile.[98] Jerome gives us the beloved hermit Paul on the front of the coin, so to speak, his own Antony on the reverse, authorizing the portrait; yet these two are neither vigorous nor virile but rather feeble and modest, not conquering the territory but disappearing into the landscape. "You see before you a man soon to become dirt," Paul proclaims (Life of Paul 10).

Never mind that royal coin-minting would scarcely have required the secrecy of a remote desert cave, however fraught and unstable the political context of Egypt might have been. In fact, the location of Paul's cave also makes it an improbable hideout for forgers, who would have been more likely to set up shop in the environs of a major urban mint. Susan Weingarten has therefore suggested that Jerome has transported forgers' hideouts from the area around Antioch, or possibly Trier, landscapes that were known to him, into the far reaches of the Egyptian desert, which was not.[99] We do, after all, tend to look for familiar features in unfamiliar landscapes, even going so far as to project them onto such scapes—just as we do in all our relationships. Whatever liberties Jerome has taken by inserting the remains of this "secret" mint into his narrative, the "scabrous" tools in Paul's cave stand as a reminder that every landscape is composed of the debris of its own history, even as it is also in a state of ongoing emergence. Every landscape carries the imprints of shifting assemblages of actors—animal, plant, and mineral. Arid landscapes in particular are often marked by human histories of economic exploitation and war, major contributors to deforestation and desertification—phenomena that in turn come to hold humans in their relentless grip. Less obvious to us, perhaps, is the impact of metallurgy. Both the forging of the iron implements and the use of those implements to mint coins require large amounts of fuel in the form of wood charcoal to refine the metal and render it malleable. Metallurgy is likewise deemed one of the major contributors to deforestation and desertification in Roman antiquity.[100]

The trees and water are gone from this particular landscape, but for a small spring and a single date palm. Most of the humans are gone too, but for two very old monks. Yet the metal shaped by the energy and efforts of trees, water, and humans still lives. I confess that my eye wants to erase it, to "clean up" the landscape, return it to its "natural" state. Yet that is once again to imagine humans as separate from nature, to construe nature as a pretty but static picture, "out there." It is also to indulge in a dangerous nostalgia, supported by a sustained practice of looking the other way. We must be willing to gaze with clear eyes upon the landscapes we traverse, whether directly or indirectly, seeing the beauty in the rusty tools and tarnished coins that mingle with the rock and sand of the austere desert.

I myself have added the coins to the picture. Jerome attributes his knowledge of the mint to written texts, "Egyptian documents." But so often it is the coins themselves that help archaeologists and historians tell time and track movement through space. So often it is the things that tell the stories.

* * *

When Zosimas first encounters Mary, they are "both running toward a place where a dry streambed had left its traces." Sophronius pauses to opine, "I do not think that a torrent ever existed there (for how could a torrent appear in that land?) but the place happened to have such a setting" (Life of Mary 11). Zosimas and Mary face each other warily from opposite banks of the wadi before approaching one another for a blessing. Later they will face each other from opposite banks of the Jordan River, a distance closed when Mary walks across the surface of the water, as if this too were a dry streambed. Later still,

Zosimas will return to the original wadi to discover Mary's corpse awaiting him on its bank. When I enter this landscape, I too approach that wadi.

Wadi is an Arabic word; it refers to "ephemeral water courses in arid regions."[101] Sophronius assumes that the wadi where Zosimas meets Mary is a trace or relic of a once flowing stream; yet at the same time he cannot imagine that any stream ever flowed in that desert. I have walked along wadis in Israel, Palestine, and Jordan and felt the same disbelief: how could water ever have flowed there at all, much less a torrent? Squinting from under my hat brim at the desert shimmering in the merciless sun, even with a backpack full of water bottles I feel parched after a few hours of hiking. So dry! How do the goats survive? The camels? The humans? Wells are the most precious of features in such a landscape: no wonder they are so fiercely guarded and sometimes fought over. Yet I have heard the stories, some tragic, of how the water can rush through these dry beds following a winter rain—in torrents, yes, sweeping unwary hikers to their deaths. Beware of flash floods, the locals say. Less dramatically, I know what it is like to find my path impeded sooner in the season than expected by water in the wadi, sending me and my companions scrambling up steep banks, searching (in my case, a bit anxiously) for handholds and footholds that will take us to higher, drier ground. I know too the delight of discovering a landscape transformed later in the season, the wadi swelling gently, edged with vivid green, at the foot of a precariously perched monastery. A paradise!

Does Zosimas intuit that Mary's sunbaked body is like a wadi? An ephemeral watercourse: there a torrent has flowed, and might flow again.

Wadis do not only channel floodwaters following the short, intense, and unpredictable rainstorms characteristic of desert climes. They also thereby allow for the replenishment or "recharge" of groundwater. Direct rainfall or smaller floods are likely to be absorbed by the soil without contributing significantly to the underlying water table. In contrast, "indirect recharge produced as a result of infiltration during flood pulses is considered as the most important contribution to the groundwater table in wadi channels."[102] Wadis distribute floodwaters across a larger terrain while also concentrating the flow, so that the water does not simply evaporate or dissipate but also penetrates the soil and reaches the groundwater below. Not all parts of the landscape are visible.

There is no recharge of groundwater without rain, however. The Jordanian desert where Mary roamed has a notoriously arid climate, with meager rainfall and a rising population, owing largely to the high number of immigrants coming from more politically troubled neighboring countries. As demand for water increases, supply will only decline: "The climate change scenarios expect a reduction of rainfall about 20 to 25% in the dry season (April–September) and 10 to 15% in wintertime . . . in Jordan during the current half of the century."[103] Thus Sophronius's failure of imagination—"how could a torrent appear in that land?"—may prove prescient.

None of us is protected from the global effects of desertification, even as the ecologies of arid regions are increasingly fragile. We must learn once again how to inhabit desert landscapes.

Holy Disfigurations: The *Life of Syncletica*

The anonymously authored fifth-century *Life of Syncletica* models its heroine in part on Athanasius's Antony: when Syncletica's parents die, she departs from her Alexandrian residence, renounces all her possessions, and heads to a tomb outside town, resolving to lead an ascetic life henceforth; whereas Antony leaves his sister in the care of others, she keeps her sister with her, a variation on a shared theme (Life of Syncletica 11).[104] Yet despite these Athanasian echoes, the text is strikingly devoid of the topographical cues that guide us through the landscape of the *Life of Antony*. Syncletica and her sister do not seem to linger long in the tomb: soon after her departure from the city, Syncletica is said to be once more in her "paternal courtyards" (13). Later, we learn that she has "withdrawn" (ἀναχωροῦσα) by herself (21) but are not told where she is practicing her solitary life or what has happened to her sister, who is blind (11). Indeed, although Syncletica is generally renowned as a "desert mother," her hagiographer shows no interest whatsoever in the desert landscape. Initially Syncletica's interior landscape looms large, as the arena or battleground for her struggles with temptation. Ultimately, however, it is her strikingly disfigured body that becomes a landscape unto itself.

Humiliating the Human

In between these two landscapes a lengthy pedagogical discourse is inserted, taking up most of the text. This "divine symposium" (30) floats free of any particular place, yet one of its key themes—humility—will be inscribed on the landscape of Syncletica's mortified flesh. "It is difficult to acquire humility," Syncletica notes, yet "it is impossible to be saved without humility" (56). She advocates meditating on scriptural passages that invite the reader to imagine herself as the lowliest and most disgusting of beings: "I am a worm and not a human being" (Ps. 21:7, LXX); "I am earth and ashes" (Gen. 18:27); "All human justice is like the rag of one who menstruates" (Isa. 64:6) (Life of Syncletica 50). Such leveling of human privilege is intended to level hierarchical distinctions between humans as well. Syncletica encourages her audience to welcome insults and even physical beatings from those they might deem their inferiors. "Humility corrects by means of reproaches, by means of violence, by means of blows," she asserts, "in order that you might hear through the foolish and the stupid, the poor person and the beggar, the weak one and the insignificant, the one who makes no progress in work, the illogical in expression, the dishonorable in appearance, the weak in power" (58). To achieve and sustain humility requires repeated acts of (self-)humiliation, so great is the human tilt toward arrogance. No one models this better than Christ: "He took the form of a slave, he was beaten, he was tortured with blows" (58).

Anti-Cosmetic Beauty

The *Life of Syncletica* does not merely argue for the necessity to submit to humiliation. It also *performs* the humiliation of the saint, first in the staging of her internal struggles and punishing austerities, then—and most dramatically—in the depiction of an illness that radically transforms her body. Like other female saints, as well as the romantic heroines they resemble, Syncletica is said to have possessed great beauty in her youth: "For she was so exceedingly beautiful physically as to attract to herself from her first youth many suitors. Some were attracted by her abundant wealth, some by the decorum of her parents; but over and above these things they were attracted to the beauty of the girl herself." Yet Syncletica is not swayed by such a flattering response to her beauty; the only bridegroom who interests her is the divine Christ (7). After her parents die, she lays aside all "cosmetics" or "adornment" (κόσμησις), going so far as to cut off her tresses (πλόκαμοι)—an act effective on its own terms, we might imagine, but also carrying symbolic freight, as our author emphasizes. "For it was the custom for women to call hair [θρίξ] 'cosmos'" (11).

Cosmos is a word with a wide semantic range, covering orders and orderings of all sorts—social, political, natural, aesthetic; in the aesthetic register, it can also indicate an ornament or decoration. Thus we are not only encouraged to read Syncletica as rejecting the particular forms of female self-adornment or self-fashioning (κόσμησις) conventionally associated with "vanity and self-indulgence," as Lynda Coon notes, in her exploration of the patristic "theology of the cosmetic."[105] We are also encouraged to read Syncletica as rejecting order itself (κόσμος). Refusing this conventional canon of beauty, she thereby becomes an "uncosm(et)ic" or disordered body—asymmetrical, disharmonious, disproportionate.

No wonder our author tells us at the beginning of the *Life* that we may have difficulty perceiving Syncletica's beauty. Like a precious pearl that is unrecognizable as such to an unpracticed eye, the "pearl" that is Syncletica does not seem extraordinary to one who considers only its form, figure, or outward appearance (σχῆμα). "But little by little from proximity we have learned its beauty [κάλλος]; divine desire [ἔρως] was born within us for what we saw; for these circumstances kindled our thought toward passion [πάθος]," the author professes (1). It is intimacy with the "strange stranger" who is our neighbor that reveals beauty. To see that beauty, we must risk familiarity. And if we do so, we will be changed. Retrained, our eyes will see differently. We will fall in love with something that initially seemed deformed or disfigured: the reader is warned.

Corpse Meditation

Yet perhaps no such warning could prepare us for Syncletica's "final struggle," which removes her beyond human comfort or cure (104). Disease first seizes her lungs, advancing gradually; bit by bit, she coughs up her very organs; fevers whittle her body away (105). She is eighty years old when her intestines are consumed with a burning pain that lasts three-and-a-half years (106). Still, she continues to teach her followers (107–9) until her vocal cords are stricken, depriving her of speech (110). Now all they can do is gaze upon her, and she upon them, but soon she becomes a sight difficult even to behold. Her teeth and gums are infected with disease, and the corruption moves into her jaw and beyond, causing the bone to fall away; within two months, all that is left of her mouth is a gaping hole surrounded by black putrefaction, and still the corruption spreads. Not only is Syncletica nearly impossible to look at—a black hole into which sight itself disappears—but she also gives off a "savage" or "inhuman" (ἀπάνθρωπος) stench. Her followers can barely stand to be near her, burning incense when they need to approach her. Initially she refuses the help of a physician but later relents for the sake of those around her, when he offers to anoint her body like a corpse with aloe, myrrh, and myrtle so as to mask the putrid smell (111). For three more months she suffers thus, unable either to eat or to sleep yet still strangely strengthened, before she is released from life (113).

Previously Syncletica has told her companions that they should counter desire-provoking fantasies of beautiful bodies as follows: "Erase the eyes of the image, and extract flesh from the cheeks; cut away under the lips, and further imagine the ugly coagulated state of bare bone. . . . For the object of love was nothing but a mixture of blood and phlegm. . . . It is necessary on the whole to represent the body of the beloved as a wound that smells oppressive, and is inclined to putrefy, briefly put, as resembling a corpse, or to imagine oneself as a corpse" (29). This exercise is strikingly similar to the Buddhist practice of corpse meditation, in which "practitioners are instructed first to become intimate with hair, nails, teeth and skin (all dead matter), then to spend as much time as possible around a dying or dead body, holding it mindfully in all its visceral gore."[106] In effect, the *Life* itself has performed the imaginative experiment with respect to Syncletica's own body, from which beauty has been progressively subtracted, leaving finally a putrefying corpse. But are we meant to love that body any less? Her last recorded words announce that she will soon "depart from the body" (113). The point of such a vivid enactment of her illness might be, then, to allow readers to detach from Syncletica's body themselves, transferring their love to her departing *soul*. Yet much in the text tugs against that conclusion. If we are meant to love her *body* less, that denigrated body should be understood as an ordered "cosmos"; loving the ordered, cosmetic body less, we are meant to love her transient, corruptible *flesh* all the more. That is the beauty learned through an intimacy that gives birth to desire and kindles passion, as evoked at the beginning of the *Life*. That is the beauty encountered in the suffering, disfigured body so spectacularly on display at the end of the *Life*. Much as with Buddhist corpse meditation, beholding the disintegration of Syncletica's body "invites us to crawl under the fence of our cultural rationalizations, categorizations, and organization of the body so as to recognize that what we find repulsive, the horror that we seize upon, is as constructed as are our notions of beauty."

Working through such affective responses, one may hope to achieve a "practiced forbearance," suggests Sharon Betcher, preparing one for compassionate encounters in the midst of mutual vulnerability.[107]

A New Aesthetics

It is not easy to perceive beauty in bodies—human or other—that are broken, diseased, aged, discolored, warped, unbalanced, askew. Yet it is increasingly difficult to ignore the fact that bodily norms do not hold, that all bodies are queer and unstable, terrifyingly vulnerable and wondrously resilient at once. If we have, in some sense, "ruined" the planet, then we must learn to love the ruins. Yet this is perhaps just an extension or intensification of the challenge of embracing the mortality of flesh as such—the challenge that Syncletica and other saints offer as they present themselves without adornment, old, sick, frail, and smelly, inviting us to discover their strength and beauty. What we think of as disabilities are "the etchings left on flesh as it encounters the world," in the words of Rosemarie Garland-Thomson. "Our bodies need care; we need assistance to live; we are fragile, limited, and pliable in the face of life itself. . . . What we call disability is perhaps the essential characteristic of being human."[108]

"Aesthetics is the science of discerning how some bodies make other bodies feel," suggests Tobin Siebers. He adds, "Art is the active site designed to explore and expand the spectrum of humanity that we will accept among us."[109] Whatever one may think of this as a general definition (it surely has limitations), I find it suggestive for this exploration of ancient *Lives* of Saints. When we encounter the saints in their literary performances, their bodies make our bodies feel a range of things that may include, for example, both disgust and admiration. If we are good readers, we may find that not just our acceptance but our desire expands—and not only in relation to a strictly human spectrum. Art, including literary art, teaches us to feel differently, taking "feel" in both the sensory and the emotional sense. Siebers also proposes that disabled bodies play a particularly strong role in modern art, referencing recent art's "love affair with misshapen and twisted bodies, stunning variety of human forms, intense representation of traumatic injury and psychological alienation, and unyielding preoccupation with wounds and tormented flesh."[110] One might add to the list, as Siebers does, the attraction of broken or fragmentary art, such as the famous Venus de Milo.[111] These are not, however, uniquely modern fascinations. Premodern Saints' *Lives* also focus attention on bodies that disrupt classical ideals of beauty, integrity, youth, and health. Bodies like Syncletica's challenge us to discover beauty where it is often obscured or denied—in the body's "relation to fragility, corporeality, pain, the animal body, the call of the other," as Betcher puts it.[112] Where pain and suffering can be looked in the eye, compassion may arise; the call of the other may be heard.

"Ecology is the latest in a series of humiliations of the human," suggests Tim Morton in a passage quoted above,[113] though perhaps we should not be too quick to generalize. It is humiliating for those of us who have attempted to master the planet to discover that we have compromised human survival as a whole. It is humiliating for those of us who have placed such a high value on autonomy to acknowledge that we need other creatures in order to thrive. It is humiliating for those of us who have dreamed of immortality to encounter our own weakness and brokenness and to accept our own transience, as individuals and as a species. It is, finally, humiliating for those of us who have invested heavily in classical ideals of beauty to confess that those very ideals have constrained our desire in an aesthetic straitjacket. But "it is impossible to be saved without humility," as Syncletica says (Life of Syncletica 56).

Saint as Posthuman Assemblage: The *Life of Simeon the Stylite*

A certain deacon climbs to the mountaintop where Simeon (c. 390–459 CE), the famous pillar-saint, dwells. On arriving, he poses the following question: "Tell me, in the name of that truth which has converted the human race itself, are you human or an incorporeal nature?" The question irritates those gathered around Simeon, but it intrigues the saint himself. He orders that a ladder be brought so that the deacon can ascend to the top of his pillar and "first examine his hands and then put his hand inside his garment of skin and see [ἰδεῖν] not only his feet but also that extremely painful ulcer [ἕλκος]" (Theodoret, Religious History 26.23).[114] What is the deacon meant to conclude about Simeon's questionable *humanity* from such an intimate bodily examination, in which sight and touch seem strangely scrambled?

This passage appears in chapter 26 of Theodoret of Cyrrhus's *Religious History*, an account of Syrian monasticism written in 444 CE, featuring ascetics both dead and still living. Simeon is among the living when Theodoret writes, yet the chapter devoted to him stands out for its hagiographical or panegyric tone.[115] As a start to unpacking our passage, we might observe that its sense depends heavily on scriptural allusion. The garment of skin (δερματίνου περιβολαίου) recalls the garments of skin (χιτῶνας δερματίνους) in which the divine creator is said to clothe the first humans after their fall from grace (Gen. 3:21), often interpreted by late ancient Greek authors as symbolic of fallen flesh itself. Thus Simeon implicitly invites the deacon to place his hand not only inside his tunic but also, at once symbolically and viscerally, inside Simeon's all-too-corruptible corporality—an open sore no less. At the same time, the scene invokes Jesus's display of his own resurrected body to uncomprehending disciples. In the Gospel of John, Thomas famously declares, "Unless I see the nail marks in his hands and put my finger where the nails were, and put my hand into his side, I will not believe" (John 20:25). Jesus's side is a wound, having been pierced by a soldier's sword (John 19:34), and he subsequently tells Thomas, "Put your finger here and see [ἴδε] my hands. Reach out your hand and put it in my side" (John 20:25, 27). Perhaps referencing Psalm 21:17 (LXX)—"they gouged my hands and feet"—the Gospel of Luke supplements wounded hands and side with feet, as the resurrected Jesus says to the disciples, "Look at [ἴδετε] my hands and my feet; see [ἴδετε] that it is I myself. Touch me and see; for a ghost [πνεῦμα] does not have flesh and bones as you see that I have" (Luke 24:39). Like both Adam and Christ, Simeon is no mere spirit but flesh and bone, Theodoret suggests through the medium of biblical allusion. The foot wound that the deacon is invited to examine—a wound haunted by other wounds, other bodily sites—is the mark of his full humanity, then. But is it not also thereby the mark of his divinity? To be *fully* human is not to be *only* human, by the logic of fifth-century Antiochene christology; and this is a passage buzzing with christological resonance.[116] Here, then, we come to the threshold of a sanctity that is more or other than human—a posthuman sanctity, if you will.

Before we cross that particular threshold, let us approach the saint from another angle. Theodoret's *Life*, like other literary *Lives* of Simeon, gives us a Simeon *with feet*. Those feet stand on top of a pillar night and day, for all to see (Religious History 26.22), so worn from standing that one of them is afflicted with an open wound, as we have just seen; it continually oozes pus, we are told (26.23). Visual depictions differ, however. In a wide range of Byz-

Figure 1. "Six Stylite Saints." Icon, date unknown. By permission of Saint Catherine's Monastery, Sinai, Egypt. Photograph courtesy of Michigan-Princeton-Alexandria Expeditions to Mount Sinai.

antine images appearing on church walls, mobile icons, pilgrim medallions, manuscript pages, and elsewhere, the iconic stylite saint, of which our Simeon is the first and most famous representative, *has no feet* (Figure 1). Human from the thighs, waist, or chest up, and typically facing the viewer frontally with arms raised in the gesture of prayer, his lower body, like that of a centaur, is other than human—in this case not horse but stone. The saint's lower body is his pillar, in short. The walled platform on top of the pillar, often chalice shaped (another strongly christological, because eucharistic, allusion), completes his lower torso, while the pillar itself extends below like a single leg, sometimes of roughly human proportions, sometimes longer. (As we shall see, Antonius's *Life of Simeon* depicts the saint standing on one foot for two years owing to a tumor on his thigh [Life of Simeon 17]—an intriguing parallel). Iconographic tradition consistently gives us the stylite as a hybrid of human and column, then. Other prostheses or accessories include the ladder that also appears in the story of the deacon. Fully human, and also fully nonhuman thing, Simeon is a cyborg of sorts—"a hybrid of machine and organism, a creature of social reality as well as a creature of fiction," as Donna Haraway famously defines it.[117] Once again, we come to the threshold of a posthuman sanctity.

The term *posthuman* is both a contested and an unstable term. *Posthuman* has relatives and rivals—*transhuman, unhuman*—as well as fellow travelers—*cyborg, monster, thing, object*. Its significance is articulated and debated within a rapidly proliferating array of often converging (and sometimes colliding) theoretical movements—actor network theory, affect theory, animal studies, assemblage theory, brain sciences, new materialisms, media theory, speculative realisms, systems theory.[118] Aided by new communications media and impelled by the pressures of ecological and political crisis, the quick pace of intellectual exchange around the posthuman is alternately exhilarating and exhausting, clarifying and confusing. Arriving fast on the heels of the human conceptualized as a socially and linguistically constructed subject, the posthuman (like other *post* concepts or intellectual movements) directly challenges its predecessor and yet also takes on many of the assumptions of the previous concept or movement of thought.

The posthuman points to a new way of thinking, then, albeit one also partly continuous with what came before. But does it, in addition, designate an ontological reality that is coming after something else, that is newly emerging? Put otherwise: Have we always been posthuman, even if we did not realize it, or are we only now becoming so? Does the posthuman (alternately, the transhuman) point toward a technologically mediated, continuously extended *telos* of human progress—an ever-emerging superhuman, perhaps? Or does the posthuman (alternately the nonhuman), on the contrary, announce a decisive decentering of the human within thought, as impelled by the pressing awareness of both the irreversible destructiveness and the undeniable limits of human agency that mark the paradoxical moment of the anthropocene? It is the latter position that draws me: we have always been posthuman but are only just learning to think that thought. As Cary Wolfe puts it, posthumanism "comes both before and after humanism: before in the sense that it names . . . the prosthetic coevolution of the human animal with the technicity of tools and external archival mechanisms (such as language and culture)." It also comes after in the sense that it "names a historical moment in which the decentering of the human by its imbrication in technical, medical, informatic, and economic networks is increasingly impossible to ignore, a historical development that points toward the necessity of new theoretical paradigms."[119] Or as Jane Bennett puts it somewhat more succinctly, "There was never a time when human agency was anything other than an interfolding network of humanity and nonhumanity; today this mingling has become harder to ignore."[120]

Premodern saints may, then, be posthuman because we have always been posthuman and also because they are, so to speak, prehuman—or perhaps better, prehumanistic. Premodern saints, and those writings, things, and performances that continue to bring them to life, do not share the full range of modern assumptions about what it means to be human; they also do not share the full range of ancient assumptions. These saints incarnate the human *otherwise*, indeed as a *becoming-other*.[121] Typically interpreted as superhuman—that is, as intensifying claims on and for humanity as a distinct, autonomous species, marked by the transcendence of its own material conditions, and destined for dominion—saints may more helpfully and richly be understood as explorers of the limits of human being. Evoking an ontology of becoming, they position the human at the shifting borders of its own otherness, its own insistent capacity for becoming something other than what it is. Put otherwise, saints expose the ongoing emergence, and submergence, of the human within the nonhuman, whether god, plant, thing, or animal.

Let me attempt to unfold this claim by contemplating a series of anecdotes culled from two Greek *Lives* of Simeon, one authored by the influential bishop Theodoret of Cyrrhus (c. 395–c. 458 CE) (whose text we have already encountered) and the other by Antonius (of whom we otherwise know nothing). Engaging Simeon through the medium of literary texts, I also want to accord the saint an agency that exceeds both the "historical" and the "literary," as usually understood. We encounter Simeon as a fluid assemblage of human, god, plant, thing, and animal. But we also encounter him as a fluid assemblage of texts, images, and devotional practices that collectively mediate his unsettling power and presence. As Glenn Peers puts it, "The Stylite saints reveal how the saint as an organism, as it were, could spread himself beyond his boundary and still stay himself."[122]

Becoming-Plant

From the start, Simeon's practices upset his fellow ascetics, creating controversy, according to Theodoret. One instance involves Simeon's use of a "rope made out of palm-leaves [σχοῖνόν . . . από φοινίκων κατεσκευασμένην]." The language of the text emphasizes the plant-based materiality of this object: not only is it explicitly said to derive from a palm, but the word for *rope* itself indicates *rush*, or something woven from rush plants, and then by further extension any woven rope or cord. And there is more: Simeon is said to have "planted [προφύσας] it on his skin," or even more literally, to have made it grow upon his skin. He does so by wrapping the rope so tightly around his naked body "as to wound [ἑλκῶσαι] the whole part which it encircled." Simeon's flesh becomes a bloody furrow, then, in which plants are rooted. After ten days, the blood begins to attract attention. One of his fellow monks inquires after its cause. "When [Simeon] said nothing was wrong, his fellow combatant forcibly stuck in his hand [τὴν χεῖρα ἐνέβαλε], discovered the cause, and informed the superior." (Later, as we have seen, Simeon will command someone else to stick his hand inside his garment so as to probe another wound.) The abbot chastises Simeon and is just barely able to detach the rope from his body. Simeon, however, refuses any treatment for his wound. His flesh remains open to its own becoming, to its participation in the swarm of being. If Simeon is becoming-plant, is the palm-leaf rope becoming-human too? Either way, the status of the human has been unsettled. So it is that the monks eventually order him to depart from their community (Religious History 26.5).

Becoming-Mountain, Becoming-Insect

In another experiment with bondage, Simeon fastens his right foot to a twenty-cubit, or thirty-foot, iron chain that is in turn attached to a large rock on the top of the mountain where he has come to dwell in solitude. His iron prosthesis roots him in place, "so that even if he wanted to he could not leave the confines" of the walled precinct that he has established on the mountaintop. Once again, his practice provokes resistance. A local bishop—one Meletius—instructs him that "right reason [γνώμης] sufficed to place rational fetters on the body." Apparently compliant, Simeon calls upon a smith to remove the chain from his leg. The process results in a revelation: under the iron is a piece of animal skin that has been wrapped around his leg and sewn shut, so that it must now be torn off; hiding in that skin are "more than twenty large bugs" that have been biting Simeon continuously, "though he could have easily . . . killed all of them." As saint, Simeon is not merely a hybrid of human, iron, rock, and mountain. He is also a human with animal skin, offering his flesh as nourishment for a swarm of hearty insects. Just as the monastic leader tries to restore Simeon's humanity by invoking health and removing the rope, so the bishop tries to restore his humanity by invoking reason, removing the chain and stripping his animal hide. But these operations are not successful. Once again, what remains is wounded flesh—so many "annoying bites." Simeon is not interested in imposing mind over matter.[123] He is interested in the transformation of flesh: his is a fully material sanctity (26.10).

Becoming-Icon

Simeon's fame spreads. His appeal is global. "For it is not only inhabitants of our part of the world who pour in," proclaims Theodoret. Ishmaelites, Persians, Armenians, Iberians, Homerites, Spaniards, Britons, Gauls—all make pilgrimage to visit the saint. But the saint travels as well, and he can be in more than one place at a time. Theodoret explains, "They say that he became so well-known in the great city of Rome that small portraits of him [εἰκόνας αὐτῷ βραχείας] were set up on a column [ἀναστηλῶσαι] at the entrances of every workplace to bring through that some protection and security to them." The use of the verb ἀναστηλόω—to set up as or on a *stēlē*—is suggestive. Just as Simeon will soon set himself up on a pillar, or *stylos*, in Theodoret's narrative, making of himself a living monument of flesh and stone, so images of him are taking their positions atop blocks of stone all across Rome, making up in number what they may seem to lack in stature.[124] A veritable army of guardians: the saint has gone viral (26.11).

Becoming-Column

Theodoret explains—and thus defends against "fault-finders"—Simeon's self-monumentalizing as a kind of flight. Paradoxically so, perhaps: the saint is at once static and mobile, heavy and light. On the one hand, his flight is away from the crowds who arrive, wanting to touch his "garments of skin [δερματίνων . . . ἱματίων]." On the other hand, his flight is not only away from but also toward: "He longs to soar to heaven and leave this earthy sojourn." The ascending columns multiply and grow: "First he had one hewn of six cubits, then one of twelve, after that one of twenty-two, and now one of thirty-six," Theodoret reports. The result? a "strangeness of spectacle" that must be biblically contextualized (and thereby normalized?) in its very oddity and novelty. Normalization does not really take: appeal to biblical oddballs—"men of God" who walk around naked or marry prostitutes, for example—arguably only intensifies Simeon's own queerness. He stands on a column: that's his thing. It's a strange thing, an ever growing human-stone erection (26.12).

Becoming-Human

It is at this point in Theodoret's narrative, when the saint has ascended his pillar and become famous as a miracle worker, that Simeon's humanity comes into question: "Are you human or an incorporeal nature?" (26.23). In truth, Simeon's humanity was already in question from the beginning. Theodoret himself has planted doubts. The story he is going to recount "surpasses human nature, and people are accustomed to measure what is said by the yardstick of what is natural." People become incredulous when they are told of "what lies outside the limits of what is natural," he repeats, but nonetheless he presses on (26.1). That measuring stick called human nature? Throw it out if you want to hear and understand my tale, he seems to say.

Admittedly, Simeon's response to the deacon's question seems to suggest that he is simply affirming his humanity, even that he is going to great lengths to do so. Stick your hand in my putrid flesh, he demands. Not a bodiless nature, then. And yet this is no simple or stable humanity either. For one thing, Simeon's wounded flesh is too closely linked to Jesus's flesh, as we have seen. For another thing, it has already been marked explicitly as transgressive and controversial. This is not the human body fettered by rationality to which Bishop Meletius tries to call Simeon back. This is open, fluid, swarming, uncontained carnality. In fact, it is much like what Gilles Deleuze and Félix Guattari call "becoming-animal."[125] The deacon asks, "Are you a human?" Simeon, in effect, asks back, "What is a human?" And the deacon seems to have learned something significant from the exchange. Theodoret concludes the anecdote thus: "The man saw and marvelled at that worst of ulcers, learnt from him that he took food, and then came down and told me all" (26.23). What is the marvel? That a human is not a bodiless nature, not even a bounded body ruled by reason, rather a kind of open frontier. Open finally to the becoming-other of its own death. A later interpolator of Theodoret's text assures his readers, "he demonstrated by his death, to those who did not believe it, that he was human" (26.28). Only clearly human in the moment that he ceases to be so, Simeon has finally perfected his performance. Becoming-human is not a flight from the always more-than-human flesh but rather a fall into its disintegrating depths.

Becoming-Worm

Theodoret is not Simeon's only biographer, as we have already noted. One Antonius, who also claims to have been a contemporary and indeed a disciple of Simeon, wrote a *Life* too. (A longer Syriac *Life* exists as well.) What Antonius lacks in sophistication or subtlety, relative to Theodoret, he makes up for in enthusiasm and graphic description. Two anecdotes in particular deserve our attention, both involving worms or maggots. The first is a version of a story we have already encountered, the incident of Simeon and the rope. There Antonius adds details not mentioned by Theodoret: for example, he explains how Simeon obtains the "little rope" (σχοινίον), stealing it from the bucket used to draw water from the well and then lying about it to the other monks. Antonius also increases the drama and intensity of the narrative. In his account, Simeon wraps the rope around his *whole* body and covers it with a tunic of *hair* (στιχάριν τρίχινον)—animal hair, presumably. Furthermore, he remains bound in the rope not for a mere ten days or more, as in Theodoret's account, but for a year or more, so that his flesh putrefies around it: "It ate into [κατέφαγεν] his flesh so that the rope was covered by the rotted flesh of the righteous man." Simeon is being devoured by the rope, then. The smell is terrible, and worse yet, "his bed was covered with worms," Antonius reports (Life of Simeon 5). This point is repeated twice more, as he relates how one of the monks finally goes to the abbot, complaining, "His bed is full of worms, and we simply cannot bear it" (6). Investigating, the abbot confirms that Simeon's bed is indeed "full of worms," and the terrible stench drives him away (7).

"Behold, the new Job!" the abbot exclaims. The comparison of Simeon to Job is apt: Job is said to have declared, "My flesh is clothed with worms and dirt" (Job 7:5). But it also sets up an ambiguous contrast. Job laments, "If I look for Sheol as my house, if I spread my bed in darkness, if I say to the putrefaction, 'you are my father' and to the worm 'my mother' or 'my sister,' where then is my hope?" (Job 17:13–15). What Job suffers passively and with distress, Simeon inflicts on himself actively, placing his hope precisely in putrefaction and worms. The abbot chastises the saint roundly, inquiring about the source of his bad smell, accusing him of threatening the discipline of the monastery, and even going so far as to suggest that he is a "phantom" and not "a true man from real parents." What then? A demon, perhaps, given away by his stench. Seeking the truth about Simeon's identity, the abbot orders him stripped (Life of Simeon 7).

This proves easier said than done. Simeon's animal hair tunic is completely fused with his putrefied flesh: the two are as one. The monks soak him in water and oil for three days; nonetheless, when they remove the garment, much of Simeon's flesh comes off with it. What remains of this human is a raw wound in which the rope is still firmly embedded. Moreover, "there was no guessing how many worms were on him." Two physicians are summoned and they labor mightily to separate the rope from Simeon; he suffers so much that "at one point they gave him up for dead," and when the rope is finally removed, flesh remains attached to it. Simeon requires fifty days of recuperative care to recover from this virtual flaying. In Antonius's handling, the saint's motivation for all of this self-imposed suffering is penance for his sins rather than training in endurance, as Theodoret implies; he refers to himself as a "stinking dog" who deserves to die. Although the abbot seems positively impressed by the theological insight that all humans are born in sin, he nonetheless orders Simeon to leave as soon as he is well enough (8). As in Theodoret's biography, Simeon is a controversial figure, deemed unfit for monastic life. He may not be a demon, but he is also no ordinary human. Even more than in Theodoret's rendering, he is an intensely hybrid figure, his flesh intimately conjoined with rope, animal hair, and worm. The abbot and the physicians try to separate the human from the vegetable and the animal, but when they do, what remains is not a contained and controlled human body but a gaping wound.

But there is more to this worm story. A second anecdote partly parallels Theodoret's account of Simeon's ulcerated foot. When Simeon has mounted his final and highest pillar, said to be forty cubits (or about sixty feet) high, he develops a tumor on his thigh, "just as happened to the blessed Job," as Antonius reports (cf. Job 2:7). As a result, Simeon has to stand on one foot for two years. Here worms again make their appearance. Antonius writes, "Such huge numbers of worms fell from his thigh to the earth that those near him had no other job but to collect them and take them back from where they had fallen, while the saint kept saying, 'Eat from what the Lord has given to you'" (Life of Simeon 17).

It is a remarkable scene, to say the least. As before, the great number of the worms is emphasized but now all the more vividly. There are so many that several of his helpers spend all of their time collecting them as they fall; one wonders how they get them back up the ladder of the sixty-foot pillar. Simeon, for his part, does not merely endure the infestation of maggots but welcomes and actively sustains it. Like Christ, he offers his flesh as food, not to other humans or even to bugs but to the lowliest—and in this case, arguably the most disgusting—of creatures. Simeon is, rather literally, becoming worm, even as the worms are becoming human. The reader is encouraged to feel not disgust but wonder and to see not monstrosity but beauty in this transformative process: Antonius reports that when a visiting king picks up one of Simeon's maggots, against the saint's initial objections, it is transformed into a pearl (18).

Becoming-Flesh

A partially gilded silver plaque from late sixth-century Syria depicts a bearded man standing on a fenced platform atop a column, visible from a little below the waist up, facing frontally and reading a book. A large serpent coils around the column, raising its head, which is depicted in profile, to gaze on the saint from the right side of the column. A ladder leans against the left side of the column, and a shell hovers over the saint's head, all in a pleasing balance. A Greek inscription at the bottom reads, "In thanks to God and to Saint Simeon, I have given" (Figure 2). We do not know who dedicated this plaque to Simeon or even whether it represents our Simeon or the younger stylite saint of the same name, nor do we know what favor the saint had granted to the donor of the plaque, obviously a person of some means. The image, however, is striking. Saint, column, ladder are familiar elements in artistic renderings, the ladder reminding us that the column saint was scarcely isolated—on the contrary. How else could his wounds be touched, his food delivered, fallen worms returned to his flesh? More unusual are the elements of book and serpent.

Antonius tells the story of a "huge dragon" who lives east of Simeon's mountain. One day a piece of wood lodges in its eye; the sound of the dragon hissing in pain can be heard far and wide. Finally, the creature comes and lies down in front of the gate to Simeon's mountaintop enclosure. Miraculously the wood falls from its eye, and the dragon stays at the holy man's gate for three days until fully healed. As Antonius marvels, "He had lain before the entrance of the righteous man just like a sheep: everyone was going in and out and nobody was hurt by him" (Life of Simeon 19). Perhaps our plaque recalls this story. The gazing serpent is, then, among other things a stand-in for the

Figure 2. "Saint Simeon Stylites." Reliquary plaque, repoussé silver with gilding, late sixth to early seventh century, Syrian, 26.9 × 25.5 cm. Louvre Museum, Paris. Photographer, Herve Lewandowski. © RMN-Grand Palais/Art Resource, New York.

plaque's donor, expressing gratitude for its healing. Or perhaps its associations are more darkly demonic and even phallic,[126] representing the temptations that beset the saint, who protects himself with the scriptures. Either way, the coiled serpent, like the column and ladder, appears an inseparable component of the holy assemblage. The saint is at once human, monument, and animal.

The saint is also silver and gold icon; the saint is also book. Simeon was not known as a reader: Antonius's abbot is surprised when he quotes scripture to good effect (Life of Simeon 8). But through his ancient biographers, Simeon himself comes to be seen in biblical figures, not least Adam, Christ, and Job, as we have noted.[127] Through his ancient biographers and all of the rest of us who write of him, he himself materializes in text. As both image and text, the stylite saint becomes a *thing*—a thing that can be repeated and replicated and venerated and beseeched, a thing with force in the world.[128]

A text might, like a rope, be made from a plant.[129] It might also be made from an animal skin. Ink is fused to the plant or animal skin surfaces. Pen-wielding humans labor on those surfaces with blood, sweat, and tears, as well as ink; they leave traces from the oils of their own skin. These elements can no longer be separated out.

Simeon wraps himself in his human, animal, plant, and mineral skins. They are inseparable. This is his becoming-flesh. It is not a defense but an exposure. It is a wound. A teeming multiplicity. Finally: it is something divine. You can reach in and touch it with your hand if you dare.

Interlude: Performance Art

Saints are performance artists, their bodies their primary medium. Their art pushes boundaries—the boundaries of art, the boundaries of performance, the boundaries of bodies, the boundaries of humanity, the boundaries of reason and sense. It makes us uncomfortable. It is supposed to.

Performance art typically values the "live," the immediate, the ephemeral; it unfolds in "real time" and can take place anywhere. It refuses commodification. It cannot be bought or owned. But at the same time, performance art blurs the distinction between the live and the virtual, the immediate and the mediated, the ephemeral and the persistent; it confounds divisions between ordinary and ritualized time and space, as well as between actor and spectator, human and nonhuman, self and other.

Celebrating its own transience, performance art also wants to leave lasting impressions: consider the artist self-baptized "Sainte Orlan," whose 1990s project *Reincarnations* involves a series of publicly performed body-altering surgeries;[130] consider too Ron Athey's blood-letting self-mutilations in "Saint Sebastian," an often-performed segment of a larger work *Martyrs and Saints* (1992) for which Athey's own HIV-positive status is by no means incidental.[131] Repetition is courted, if only inevitably to be thwarted: in her 2005 *Seven Easy Pieces*, staged at New York's Guggenheim Museum, the famous Marina Abromovic reenacts a series of influential performance works from the 1960s and 1970s, thereby exploring "the possibility of redoing and preserving an art form that is, by nature, ephemeral."[132] Photographs and videotapes of performances provide another medium of preservation, easily disseminated via the internet: the 2013 short film *American Reflexxx*, which follows performance artist Signe Pierce walking down a busy street in stripper garb and a reflective mask while onlookers denigrate and eventually attack her, had over a million-and-a-half views within two years of being published on YouTube in 2015.[133] When asked if there is "a strong dividing line" between live and mediated formats in her own work, performance artist Helen Paris replies, "Not really. It comes back to contact and communication. . . . Use whatever lets you tell the story best." She adds, "The relationship between live and mediated formats in performance fascinates me, the visceral and the virtual—the immortality and mortality of both forms."[134] Writing is of course a form of mediation as well.

Mortality itself is often the explicit subject of performance art as a cultivated transience is tensively balanced against improbable feats of endurance and persistence. Gina Pane's *Death Control* (1974), recorded in both photographic and video formats, offers a close-up shot of live maggots crawling across the artist's face, which remains completely still (Figure 3). Pane recounts, "I was living in posthumous time. Covered with maggots, my flesh detached by maggots: flesh of my flesh, two fleshes living together, one nourishing itself from the other: process of life in a continuum of time."[135] A reviewer comments that the videotaped performance is "unbearable" to look at: the maggots "wriggle in her ears and appear to try and get in under her eyelid. The obvious association with death and disease is disturbing—we know flies land on corpses and lay eggs, which develop into maggots, and that

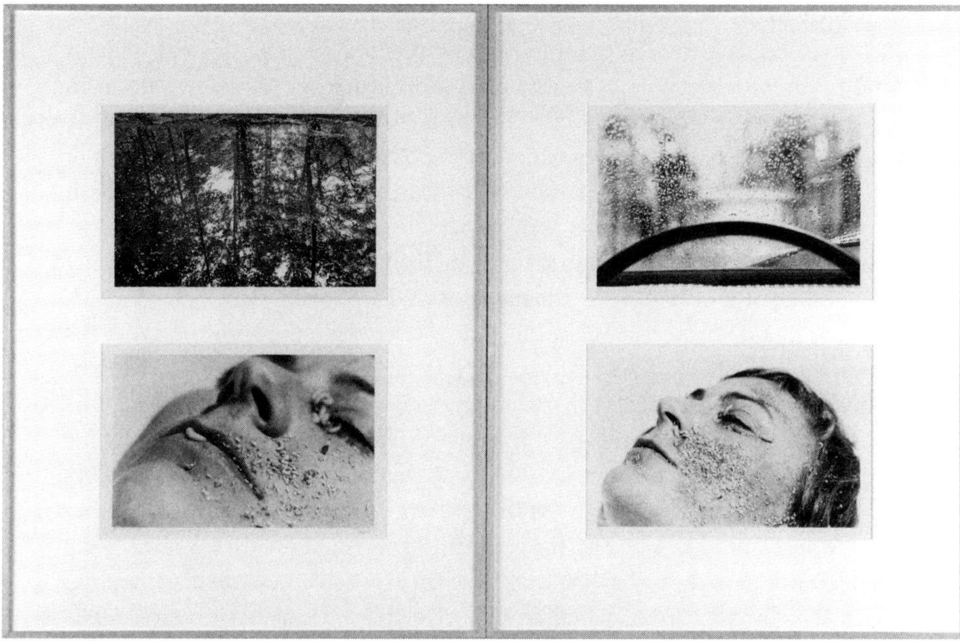

Figure 3. Gina Pane. "Action posthume de l'action Death Control, 1974." 81 × 80 cm. From the collection of Anne Marchand. Photograph courtesy of Frac des Pays de la Loire. Photographer, Cécile Clos/Ville de Nantes, Musée des Beaux-Arts. © 2018 Artists Rights Society (ARS), New York/ADAGP, Paris.

they are generally considered vile, disgusting creatures which will eventually burst forth as blowflies."[136] Like the literary *Lives* of Syncletica and Simeon, the videotape of Pane confronts spectators with the unsettling porosity and corruptibility of mortal flesh, evoking a range of possible feelings—horror, disgust, grief, fascination, awe. Crucially, death is interwoven with birth—"process of life in a continuum of time." In fact, Pane's installation includes a second film that is displayed simultaneously: alongside the video of maggot-covered flesh runs a video of a birthday party at which children are destroying a cake. The work explores "the juxtaposition of two kinds of time," as Pane puts it, "the time of death and the day we celebrate every year with a party which marks the approach of death."[137] The ruined cake echoes the ongoing disintegration of the body.

But mortality does not only haunt natality: natality also haunts mortality, as flesh nourishes flesh. Another artist, Teresa Murak, uses the warmth of her own body to germinate plant seeds. As a student in Warsaw in the early 1970s, she sowed seeds into her clothing in order to stimulate their growth—a faint echo of Simeon planting rushes in the bloody furrows of his own flesh. In subsequent performances of *Seed* in Berlin (1989) and New York (1991), Murak lay in a bathtub full of wet seeds for several days, again using the warmth of her body to stimulate germination and thus deliberately weaving her flesh into an unfurling tapestry of cosmic growth.[138]

Still other artists have gone so far as to display their own process of dying in the midst of mortal illness. The notoriously flamboyant Bob Flanagan participated in a documentary about his life and work on the condition that the film include a record of his death from cystic fibrosis, a disease that, like Syncletica's, particularly affects the lungs, producing terrible fits of coughing; the film concludes with a series of still shots of Flanagan's corpse, taken by his longtime partner and fellow artist, Sheree Rose.[139] Hannah Wilke's work offers even more striking resonances with Syncletica's performance. Early in her life, viewers found her beauty both alluring and (in some cases) disturbing: her use of her body in her art making was especially controversial among feminists who questioned whether her self-display critiqued or merely replicated pornographic images formed by an objectifying male gaze. Wilke's performances included covering her naked body with tiny figures of female genitalia molded from chewed bubble gum—an act performed live, using gum chewed by the spectators, but also represented photographically as the *S.O.S. Starification Object Series*. Yet Wilke's critics as well as her admirers came to understand the trajectory of her life work differently from the vantage point of the punningly titled *Intra-Venus*, created in collaboration with Donald Goddard. Her final project, this series of posthumously published photographic and watercolor self-portraits and other related objects displays the body of a middle-aged woman ravaged by cancer and its treatments—hairless, swollen, bruised, bandaged, depleted. Amelia Jones comments as follows regarding one of the images, "Intra-Venus Series No. 10" (1992), a photograph taken about a year before Wilke's death: "The beauty here is not that of appearance, but of being—a being that persists, struggles, in the face of death's inexorable and 'untimely' approach."[140]

That "being" only arrives in and as its own appearance, however, an appearance rendered all the more vivid and poignant by the awareness that life's persistence—its transcending beauty—always manifests in the face of impending dissolution. In "Intra-Venus Series Number 4" (July 26, 1992),

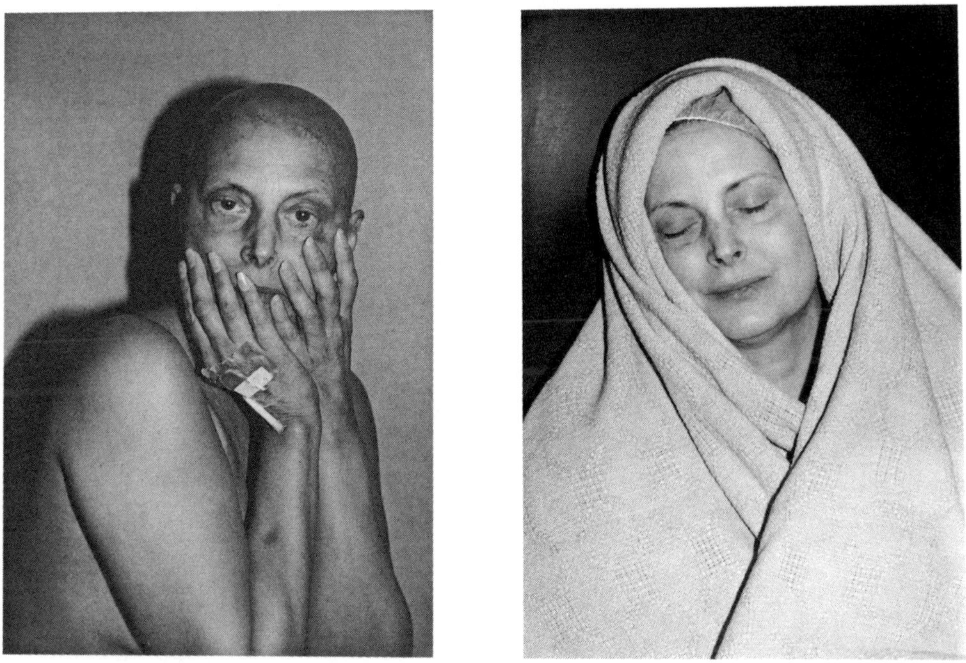

Figure 4. Hannah Wilke. "Intra-Venus Series Number 4, July 26 and February 19, 1992–1993." Performalist Self-Portrait with Donald Goddard. Two chromogenic supergloss prints, 71½ × 47½ inches each. Courtesy of Donald and Hellen Goddard and Ronald Feldman Fine Arts, New York.

Wilke's gently tilted head is swathed in a pale blue hospital blanket; her eyes are shut; and the ghost of a smile haunts her closed lips (Figure 4). She "looks like nothing but a Madonna—yet under and around the eyes—an ominous purplish cast points toward trouble," writes Jones. "The smile, on second glance, is tired rather than inspired. The face glows, but perhaps the glow is otherworldly."[141] "Otherwordly" is precisely what we would expect from a Madonna. Why, then, the warning "but perhaps" that aligns it with the bruised and the exhausted, making it one more marker of death's proximity? What is the *other world* that shines through a body that is not long for *this* world? Is it perhaps "life in process," a fleshly mesh in which our (more or less) discrete bodies are only ever temporarily suspended, from which they arise and to which they return?

In/Conclusion: Saints and Other Queer Creatures

Hagiography is the art of writing holy *Lives* and of writing life as holy. Put simply, it makes saints out of humans. Saints are both more and less than human: they are desirable, desiring living beings, displayed in all their intensity and vividness—animals, in short. They are also both angelic and demonic—wild animals, then, pulsing with mysterious powers. It is hard to say how far the boundaries of the living—of the animal or the animate—extend. It is hard to say what, if anything, is beyond the realm of life, love, and mystery—what, if anything, is beyond sanctification. Hagiography pushes at those boundaries, wherever they are drawn.

What is the relation of a living saint to a literary *Life*? Each exceeds the other. A living saint is not a mere historical personage. A living saint is the mobile and collective effect of an assemblage of performatively embodied conjurings or evocations—narrative, poetic, musical, visual, tactile, kinetic, and more. For its part, a literary *Life* (or any other artistic conjuring) gives us not a saint *tout court* but a saint enmeshed in a particular world—temporal, spatial, relational, and more. It gives us an entire (if never finished or closed) creation. Each saint and each literary *Life* is distinctive, yet it is possible to venture some generalizations.

The *Life* of a saint breaks with the chronologies inscribed by both biographical and novelistic conventions. Birth does not anchor a social identity: its circumstances are unknown, ignored, transformed, or abandoned. Nor does the *Life* follow a recognizable course of advance through childhood and youth to an adulthood shaped by marriage, childbearing, and the shouldering of economic and civic responsibilities. The saint does not advance at all, according to such human measures. Rather, the saint pauses. The saint pauses to recalibrate, seeking to synchronize with rhythms at once divine and cosmic, while refusing the pleasures, privileges, and burdens of conventional adulthood altogether. Time, for the saint, is unhurried, open, meandering, without fixed end or goal. It is a time of attentiveness and improvisation. It is the time of the very old or the very young, the time of lions panting in the shade, of slow-growing palm trees, rusting iron, dry streambeds anticipating a flood. It is the time of memory and dream, the time of encounter, the time of waiting for love's always unexpected arrival. It is the time of birth and death, the time of our ongoing birthing and dying. It is animal time.

"How little of the movements of the bodies of octopuses frolicking over the reef, of guppies fluttering in the slow currents of the Amazon, of black cockatoos fluttering their acrobatics in the vines of the rain forest, of terns of the species *Sterna paradisaea* scrolling up all the latitudes of the planet from Antarctica to the Arctics, of humans, are teleological! How little of these movements are programmed by an advance representation of a goal, a result to be acquired or produced, a final state! All these movements do not get their meaning from an outside referent envisioned from the start, and do not get their direction from an end point, a goal or result. Without theme, climax, or denouement, they extend from the middle, they are spreadings of duration."[142]

The saint does not stand alone but is enfolded in landscapes of relations that are intimately known and loved and also always unfurling, unbounded, mysterious, inviting exploration and discovery. If the landscape is a deserted place or wilderness, it is not a wilderness that is opposed to a civilized or urban place, as nature to culture. The desert itself is an assembly or community of humans and others, ever shaped by the histories of those who have inhabited and traversed it. But it is not a place dominated by humans, nor is it a place that can be imagined to exist exclusively or even primarily for humans. Rather, it is a place that can just barely support human life, alongside other life forms, some of which are distinctly hostile to humans. It is thus a place—any place—that calls humans back into sharp awareness of their animal bodies—the need for water, food, shelter from the heat and cold, protection from predators, and the ability to move across difficult terrain, for example. The desert elicits an expansive attunement of the senses and the imagination that is only possible when the clamor of (other) humans recedes. "As long as animals are granted presence in the landscape, there is still a humming to be heard."[143] The humming landscape of the desert evokes a sensitivity to beauty that is not lavish but austere and withholding, not gentle but jagged and harsh, not scaled to the human but extending and enduring beyond human experience or understanding. Then again, austerity might be the very condition for lavish display: "Desert flora are sparse and ephemeral," writes naturalist Ellen Meloy. "There are spines, thorns, uncertain seeds, long periods of dormancy, and, when moisture comes, a passion so accelerated, you feel their demands on your heart, the mounting pleasure, the sweet exhaustion."[144] A place of extremes, the desert forms a matrix of relations among queerly adaptable creatures, whether animal, plant, or mineral; saints are among them. Some of them live in caves. Some of them roam the wadis. Some of them perch on top of rocks or pillars. They do not mind the howling hyenas or the biting bugs. They may seem a bit mad.

Saints probe the limits of spiritual possibility while also plumbing the depths of physical abjection. They do not shrink from the evidence of their own mortality; indeed, they positively flaunt it. Hair and skin is unkempt, clothing rots and drops away, gender and even species becomes indistinct. Emaciated limbs grow weak. Teeth are ground down to the jawbone. A jawbone itself becomes diseased and disintegrates. Cancer spreads. Infection festers. Maggots swarm on putrid flesh that has been rubbed raw by chains and fused to bindings of cloth and rope. Dust finally mingles with dust. Life in process: the pain can be excruciating, the sight appalling, the stench unbearable.

Yet others perceive beauty, feel desire, extend compassion and care. Confronted with both the undefended vulnerability and the resilient endurance of the saint's body, they discover a version of the same in themselves. To imitate a saint is not to model oneself on that saint, as we usually imagine it. Rather, it is to respond in kind—each of us in our own way, with infinite variety—to another creature's poignant fragility and astonishing persistence. As the novelist Dorothy Allison writes, "I wear my skin only as thin as I have to, armor myself only as much as seems absolutely necessary. I try to live naked in the world, unashamed even under attack, unafraid even though I know how much there is to fear. . . . I have to try constantly to understand more, love more fully, go more naked in order to make others as safe as I myself want to be."[145]

These are "ecology's unfathomable intimacies," toward which Timothy Morton gestures.[146]

Hagiography makes saints out of humans. It is a form not of biography but of zoography. That is to say, it does not chart the impressive career of an exceptional human being but celebrates the liveliness of creaturely existence in which all humans participate—albeit some more intensively, more wholeheartedly, than others. It is those who live most intensely, most wholeheartedly, that we call saints. On this reading, hagiography undoes human exceptionalism. In the same gesture, it undermines any concept of natural order. The behavior of saints is unnatural, flamboyantly so. Beholding the saints, we perceive that there is no nature as such, only shifting landscapes of mutually enfolding, ever emerging creatures, strange strangers all.

Does that mean that all forms and states of creaturely existence are equal? If there is no normal or natural to oppose to the queer, no healthy or able-bodied to oppose to the crip, do distinctions cease to matter? Not at all. The encounter with saints suggests that creatures are to be embraced in their wild and unpredictable diversity, a diversity that queers and thus broadens our given human ways of knowing, loving, and living. "We carry," asserts Betcher, "our most intimate view of nature within our expectations for a given state of health." "What if we crips too claim to be 'the picture of health'?" she asks. What if health "refuses to resent the world"? What if it comes in more than one guise? What if evolution is not teleologically driven but open ended? She proposes, "Physical and/or neurological alterity might even then be lived as creative mutation, spiritual teacher, double consciouness, creative assemblage with neighbor or even earthy elementals."[147] The queer and the crip—the saints among us—may be our most precious guides for living creatively into a precarious and unpredictable future.

They are also the most strongly marked bearers of social and fleshly vulnerability, reminders of the high costs of human injustice built on narrowly defined ideals of the normal, the natural, the desirable, the beautiful. Yet saints do not inspire our pity but rather our awe. They challenge us to armor ourselves only as much as we absolutely have to, granting others the safety that we ourselves desire. They challenge us to live responsively and courageously in the midst of shared vulnerability. They challenge us to love promiscuously.

Postlude: A Tough Love

Some creatures are easy to love. Most are not.

It is tough to love a saint.

A tough love. A queer love.

So much flesh! Theirs. Mine.

I have always had a strong gag reflex. The sight of blood, pus, anything piercing the skin makes me feel nauseous and faint. Maggots—forget it. I am also afraid of heights.

Yes, the thought of so much mortal flesh makes me a bit dizzy.

And what could we possibly talk about? Religion? Sex? The future of the human race?

Yet they draw me too, those queer old saints. Not despite but because of all that.

Perhaps we would not do much talking. It is easy to sit quietly in the desert, listening to the faint hum of the heat in your ears, the rustle of a lizard, or the buzz of a fly. It is easy to sit quietly in the desert, gazing upon the sun-drenched sand and rock, the dusty green scrub, the unexpected brightness of a blossoming flower.

Perhaps, over time, we would grow comfortable with each other's animal bodies, with how the other's animal body made our own animal body feel. Perhaps I would not gag as easily. Or I would mind my gagging less. My strangeness would become more familiar to them as well. We would not do much talking but we would risk the comfort of touch.

One day I would awaken to discover that the saint had died in the night. I would lie down beside the body and hold it in my embrace, realizing that I was no longer afraid of death. Or no longer afraid of my own fear.

Perhaps, over time, I would learn to see my own demons, grow my own vegetables, gather strange desert fruits, drink water from a spring as if it were wine. Live in a cave, roam the wadis, perch in high places. Grow my hair, shed my clothes. See the beauty in the rusting carcass of a pickup truck half buried in the dirt. Listen to the call of the other animals. Read and write messages in the sand.

III

Things and Practices

Arts of Coexistence

Prelude: Theorizing Things

Why advocate the vitality of matter? Because my hunch is that the image of dead or thoroughly instrumentalized matter feeds human hubris and our earth-destroying fantasies of conquest and consumption. It does so by preventing us from detecting (seeing, hearing, smelling, tasting, feeling) a fuller range of the nonhuman powers circulating around and within human bodies. These material powers, which can aid or destroy, enrich or disable, ennoble or degrade us, in any case call for our attentiveness, or even "respect."

—Jane Bennett

Things have received a great deal of scholarly attention in recent decades. They were already a going concern when Bill Brown's influential 2001 essay "Thing Theory" marked the theorizing of things as . . . well, a *thing*.[1] Indeed, some might say that thing theory was well launched by 1950, when Martin Heidegger delivered his lecture "Das Ding" in Munich. The questions the German philosopher asks are profound in their simplicity: "What in the thing is thingly? What is the thing in itself?"[2] Attempting to push beyond the deadening experience of things as the mere objects of human representation, explanation, or mastery, Heidegger proposes that things, in their true essence, are active (a "thinging" as much as a "thing"), relational (a "nearing" and a "calling"), and elusive ("nearness preserves farness," as he puts it, with sound-bite conciseness).[3] In Graham Harman's view (a self-avowedly idiosyncratic one), "The thing is Heidegger's most important idea, one that encapsulates all the insights of his long career"; Heidegger "revolutionized our concept of things."[4]

For Heidegger, however, a line is firmly drawn between things and humans, even as the relationship between them is privileged: things as such manifest only to (enlightened) humans, and humans are not themselves things. "Men alone, as mortals, by dwelling attain to the world as world. Only what conjoins itself out of world becomes a thing."[5] Heidegger's thought is strongly anthropocentric, to put it very simply. Only humans have a world, and only that world sustains things as things—active, relational, elusive. Elsewhere Heidegger makes it explicit that nonhuman animals relate to other entities as mere utilitarian objects, never as things in themselves: animals do not "dwell"; animals lack a sense of both world and things.[6] Certainly, things themselves lack a sense of world and things as well, in Heidegger's view.

What might it mean to conceptualize *things* from a nonanthropocentric perspective? This question has gripped some recent theorists, impelled in part by the pressing urgencies of the ecological crisis. After all, anthropocentrism comes under particular scrutiny "when we are forced to recognize that the fate of humanity is deeply intertwined with the fates of all sorts of other entities," as Steven Shaviro notes.[7] To this end, Harman and other self-proclaimed "object-oriented ontologists" have sought to refashion Heidegger's phenomenology.[8] The privileging of the human—or of *Dasein*, in Heidegger's terms—is not crucial to his theory of things, it is argued. Indeed, suggests Harman, when Heidegger is read rightly, "the lingering priority of Dasein in his thinking is vaporized."[9] Thus "radicalizing Heidegger," as Shaviro puts it,[10] Harman argues that all entities (including humans) are *things* or *objects* (the latter his own preferred term), and also that all entities (including nonhumans) also possess the characteristics of *Dasein* in the ability to relate to (other) things both as distinct entities and as constituting a world. It is not only a human who can relate to a rock as a rock; so too can an ant or rain. Yet human, ant, and rain will all relate to a rock differently, and none will be able to perceive (or "prehend") the rock fully, as it "really" is. For Harman, as for Heidegger, there is a tension between things as revealed or perceived and things in their withdrawn concealment; there is likewise a tension between things as relationally constituted and things as separate, independent entities. Without letting go of either tension, Harman is inclined to emphasize both the withdrawnness and the separateness of things. His things or objects are very persistently thingly, in other words, resisting absorption by systems, processes, or relations; they also sparkle with mystery in their agential forcefulness, on the one hand, and ultimate unknowability, on the other. Like other object-oriented ontologists, Harman delights in lists that suggestively perform the eclectic individuality and lively energy of things: "dust and cinderblocks and shafts of sunlight"; "person, hammer, chandelier, insect"; "fireworks, grasshoppers, moonbeams, and wood"; "hammers, roads, propane tanks, eagles, cobras, rolling woodlands"; and so on.[11]

Timothy Morton has been the most vociferous advocate of the significance of object-oriented ontology for ecological thought.[12] Decisively decentering the human, leveling hierarchies of being, and refusing the distinction between the natural and the unnatural—these are all conceptual moves that radicalize the recognition that "everything is interconnected," the crucial ecological insight touted by Morton in previous work.[13] In addition, emphasis on discrete objects honors the distinctness of all things, while emphasis on their withdrawnness underwrites their alluring and alarming strangeness: we coexist amid a multitude of "strange strangers," as Morton likes to put it. The effect of these insights, if taken seriously, is to transform a habituated sense of distance and mastery to one of intimacy and shared vulnerability. But some strange strangers are more easily spotted than others: Morton is highly attuned to differences of scale among objects and the perceptual challenges that these present. With this in mind, he has coined the term *hyperobject* to capture phenomena, such as global warming, for which "such gigantic scales are involved—or rather such knotty relationships between gigantic and intimate scales—" that they cannot be easily mapped onto humanly conceivable time and space.[14] The very small may also perplex us: what does it mean to relate intimately to the bacteria in my own gut? Are they me or not-me? Is global warming—in relation to which I am like the bacteria in the gut—me or not-me, for that matter?

Such complex yet also mundane "interobjectivity"—what Morton also refers to as "the mesh"—invites not only conceptual but also (and perhaps more notably) aesthetic and affective engagement. Classical ideals and embodiments of beauty, involving harmony, symmetry, or proportion, for example, offer lamentably inadequate cues for perceiving and relating to a universe of eclectic and uncanny things in their messy and unpredictable entanglements, of which we ourselves are a part. Even if harmony prevailed, we would have no place to stand from which to survey it with measured appreciation. The swarm of strange strangers is in our faces (as well as our guts): it always was, but now we begin to have no choice but to notice it. As a result, our senses may feel overloaded and so too our emotions. More than ever, wonder may find itself entwined with horror, curiosity with disgust, joy with grief. Such ambivalent intensities are challenging. They may also be positively transformative.

In place of "beauty," Morton proposes an object-oriented and decisively non-Kantian "sublime" that resonates with the sublime of the ancient literary critic conventionally known as Longinus. He focuses in particular on Longinus's elevation of *ekphrasis*, or vivid, intensifying description—the artistic exposure of things in their thingliness, one might say. Whereas Kant associates the sublime with an encounter with what is vast and boundless, overpowering comprehension and inspiring fearful awe, Longinus, as Morton reads him, "is talking about intimacy with an alien presence: the sublime is what evokes this proximity of the alien."[15] Such intimacy is queerly erotic and also, suggests Morton, characteristically melancholic. "The more the ekphrasis zaps us, the more we fall back into the gravity well of melancholy," he asserts. "The inward, withdrawn, operationally closed mood called melancholy is something we shake off at our peril in these dark ecological times." Why? Because melancholy bears witness to and also embodies the ungraspability of objects, their essential elusiveness or withdrawnness in relation to one another; it draws us into the "twilight shadows" of ecological coexistence itself.[16] Yet the affective palate of those attentive to the dynamics of coexistence is not monochromatic, of course: "In terms of how much they open us to ecological thought," writes Morton elsewhere, "I'd rank compassion, curiosity, humility, openness, sadness, and tenderness the highest."[17] Undergirding and undergirded by these affective states is a practice of patient being-with, a kind of "rest" in attunement—"mindfulness, awareness, simple letting-be."[18]

Jane Bennett, cited in the epigraph above, shares Morton's emphasis on the significance of "aesthetic-affective styles" for the formation and transformation of ecological desire and action: "If a set of moral principles is actually to be lived out, the right mood or landscape of affect has to be in place." Where Morton highlights melancholy, Bennett touts wonder or enchantment—"that strange combination of delight and disturbance."[19] Her objects, things, or bodies (she prefers the latter two terms) are less radically withdrawn from one another than Morton's, we may infer.[20] Extending the concept of affect beyond the human to refer to "the capacity of any body for activity and responsiveness," she equates this capacity with materiality itself.[21] "Thing-power" is one of her names for the inherent liveliness or vitality of material bodies. Yet at the same time she is wary of any conception of things as static or individualistic: "an actant never really acts alone"; rather, agency is distributed within ever-emerging (and ever-dispersing) collectivities or assemblages.[22] Agential collectivities that traverse the human and nonhuman are of particular interest to Bennett: "There was never a time when human agency was anything other than an interfolding network of humanity and nonhumanity; today this mingling has become harder to ignore."[23]

Yet for a long time, it has been ignored. As Annabel Wharton notes with respect to modernity, "Industrialization and capitalism have made most objects into fungible commodities, eliminating their particularities, their histories, and their ability to act." She adds, "Nevertheless, certain classes of things, like relics, totems, fetishes, and idols, continue to be recognized by at least some audiences as agents."[24] My own question is this: What enables some "audiences"—not just modern but premodern ones too—to recognize, and even experience vividly, "the gravity of things, their magnetic pulls," as Donovan Schaefer puts it,[25] where others do not? How, in particular, do late ancient Christians cultivate relationships to nonhuman things?

Theories of things imply *practices of things*. (Theorizing is one of them.) These practices are situated within and also give rise to "interfolding network[s] of humanity and nonhumanity," as Bennett puts it. Here I want to explore the specific practices that bind late ancient Christians in intimate relation to relics and icons, to buildings and landscapes, to rhetorically evoked and imaginally contemplated things, in all of which they acknowledge a lively agency. How are these relationships between humans and nonhuman things ritualized and enacted? How do sensation and affect circulate among them? What, finally, might such strange and distant ways of doing and feeling teach *us*, as we seek to recover a sense of the vitality and interconnectivity of things in this moment of ecological precarity?

Feeling Things: Relics and Icons in an Animate World

Let us start with a *particular* thing—a ring. This ring once belonged to Macrina, older sister of the Cappadocian theologians Basil of Caesarea and Gregory of Nyssa, memorialized in a biography penned by the latter. The ring passed from Macrina to Gregory upon her death; in fact, the *Life of Macrina* serves, among other things, to legitimate this inheritance. We do not know what happened to the ring subsequently. What we do know about it—and it is not much, really—derives exclusively from Gregory's text. This ring is foremost a literary thing, then; it is even possible that Gregory invented it. But it is no less thingly for that.[26]

Crafted from iron (Life of Macrina 30),[27] the ring imposes itself on us in its materiality. The metal is a humble one, iron being "found in the greatest abundance of all metals," as Pliny the Elder testifies; notoriously, it can serve both the best and worst purposes as a tool of either agriculture or warfare, while its great strength is mitigated by its vulnerability to rust or "bleed," lending it the poignancy of mortality (Natural History 34.39–46).[28] Some credit iron with apotropaic powers: a character in one of Lucian of Samosata's dialogues claims that a "ring made of the iron of crosses," given to him by an Arab, protects him from demonic attacks (Lover of Lies 17),[29] and the empress Helena is said to have had a horse bridle fashioned for her son Constantine from one of the iron nails of Christ's cross; she had another such nail placed in Constantine's diadem to guarantee the safety of both emperor and empire (Ambrose, Funeral Oration for Theodosius).[30] The metal is less often used for adornment, but Pliny notes with approval that the ancient Romans eschewed luxury and wore iron rings rather than gold ones, except when engaged in public affairs of state. "This is the reason why even now an iron ring, and what is more a ring without any stone in it, is sent as a gift to a woman when betrothed," he adds, with reference to first-century practice (Natural History 33.4). By the fourth century, fashion had changed, and women of Macrina's status and wealth commonly wore gold rings. Nonetheless, the custom of giving a simple iron betrothal ring may have continued, even if no positive evidence of this survives: with respect to material remains, such betrothal rings would not be distinguishable from others.[31] If earlier Roman practice is suggestive, so too is later Byzantine custom. When marriage rings for both partners come into use, a man receives a gold ring, a woman one of silver, iron, or copper. The fifteenth-century theologian Symeon of Thessalonica proposes that the iron ring symbolizes the firmness or strength that the groom pledges to his bride.[32] Thus, the iron of Macrina's ring potentially performs a range of actions—refusing extravagance, warding off evil, and binding a bridegroom to his bride.

Who is Macrina's bridegroom? None other than Christ. The betrothal is a secret one, and Macrina does not wear the ring on her finger where others might see it. Indeed, Gregory first becomes aware of its existence when he is preparing his sister's body for burial. Macrina's companion Vetiana assists him, calling his attention to the delicate chain that encircles Macrina's neck. The necklace holds two tokens next to her heart—"an iron cross and a ring of the same material," as Gregory describes them succinctly (Life of Macrina 30). He graciously offers Vetiana the cross, a *phylakterion,* or amulet (as he calls it), which he perceives to be the more valuable object. Taking the ring for himself, he observes that "a cross is engraved on the seal." But there is more to this signet ring than meets the eye. Vetiana explains, "The ring is hollowed out and in it is hidden a piece of the wood of life. And thus the seal of the cross on the outside testifies by its form to what is inside" (30). Engraved iron and living wood: a visible image calls forth a hidden presence. Macrina's ring both signifies and encloses a fragment of matter that gives life because a life has been given; a tree hewn from its root overflows with the vitality of blood spilled by Christ. This ring is, then, at once seal, icon, and reliquary. In its surplus, it takes on the character of what Roland Barthes calls a *punctum*—a small detail that nonetheless "rises from the scene, shoots out of it like an arrow, and pierces me."[33]

How might we continue to engage the ring in its piercing materiality? A striking object from the sixth or seventh century offers a suggestive comparison. A ring made not of iron but of gold, it features a polished rectangular garnet set in a grooved box frame adorned with a border of granulated pyramids and supported by an inverted cone that is soldered to the ring itself. A small, curved, six-pointed calyx-shaped container is attached to the side of this central setting, covered by a hinged gold foil cross; the hinge abuts the central setting so that when the delicate cruciform lid is open, it rests on top of the garnet stone. As Lieselotte Kötzsche-Breitenbruch notes, at least two features of this ring—the hollow container that appears to function as a reliquary and the image of a cross on its lid—recall Macrina's own ring.[34] However, the differences are perhaps as striking as the similarities. Kötzsche-Breitenbruch's descriptions of the Byzantine ring are telling: recalling the density of much ancient *ekphrasis*, they depict an artistic object so elaborately wrought that it at once seduces and eludes a verbally mediated visual imagination.[35] In contrast, Vetiana's words, as Gregory renders them, are pared down to the bone. We lack details precisely where the modern descriptions of the Byzantine ring overwhelm us. Is the seal hinged so that it can be opened to reveal the wood fragment or is it soldered shut? Is the hollow container solid or is it (as with the Byzantine ring) perforated to allow for some communication between interior and exterior? Gregory does not merely give us unadorned iron of apparently simple design where the Byzantine ring gives gold, garnet, and complex structure. His words focus attention on the ring—the *thing*—not as visual spectacle but rather in its most interactive and tactile materiality.

The iron of Macrina's ring has been impinged upon, carved with an image of the cross. Unlike the more elaborately decorated but less functional gold one, it may in turn create an imprint on wax: thus will the sign of the cross be replicated again and again. In this respect and others, the ring becomes a mirror for Macrina and she for it. She breathes her last breath as she is making the sign of the cross; previously she has made the same sign on her own eyes, mouth, and heart, charged sites of a deeply embodied piety. Like the ring, Macrina has been imprinted by the cross—indeed, self-imprinted. Also like the ring, Macrina makes quite an impression on others, not least her brother Gregory. "I too have been crucified with you," Macrina calls to her divine bridegroom from her perch on an imaginary Golgotha (24). Crossed and crucified, by gesture and word she draws as close as she can to her god, conforming her body to his and thereby positively courting its violent disruption.

Macrina has made herself not only an imprinted and imprinting seal but also an icon and reliquary for the cruciform objects that lie hidden beneath her clothing, marking the crossroads of her own heart. With arms outstretched in prayer like the *orans* figure so familiar in Christian iconography—or, alternately, with arms stretched on an imaginary cross, as her own words suggest—her body "testifies by its form to what is inside," to borrow Vetiana's phrasing. Thus Macrina's worshipping, crucified body is assimilated to the metal and wood things that it not only signals but also frames and contains, even as it is also in some sense signaled, framed, and contained by those things (Figure 5). In such a context, "the definition of inside and outside is no longer firm and fast," as the borders between saint, objects, and viewer or reader thin out.[36] At the same time, a queer leveling of scale and kind may take place across these otherwise disparate bodies and things through their shared and mutual indexicality.

Finally, Macrina is not only a reliquary but also a relic, a hidden presence signaled and summoned by Gregory's biography, itself at once icon and frame. Anticipating his visit to his sister, Gregory sees a vision as he sleeps, and he sees it not once but three times. "I seemed to be carrying the relics of martyrs in my hand," he recalls, "and a light seemed to come from them, as happens when the sun is reflected on a bright mirror so that the eye is dazzled by the brilliance of the beam" (15). These relics are presumably bones, bones that blaze like a mirror reflecting sunlight. "Reflected light is a sign of divine presence that infuses and transfigures the person before it," as Glenn Peers suggests.[37] Gregory is already bedazzled, then; his transfiguration is under way. He carries the shining fragments in his hand, a small and intimate cargo:

Figure 5. "Crosses, Byzantine." Bronze and soapstone, height 1 to 4 inches (2.5 to 10.2 cm). Photographer, Paul Hester. Courtesy of the Menil Collection, Houston.

both more and less than human, they are all that will have remained of Macrina. This anticipatory vision returns to Gregory following Macrina's death, moreover. Prepared for burial, her body glows: "as in the vision of my dream, rays seemed to be shining forth from her loveliness," he writes (32). Dead, the saint is more vividly present than ever. On the verge of disintegrating into "a disgusting and disagreeable formlessness," as Gregory describes the corpses of his parents (35), Macrina already gleams with the beauty of remains purified by time. Bits of bone and wood may finally claim an alluring vitality that vies with that of living flesh or tree. Is this because they have a complex history? Yes. But it is also because they have a densely relational present and an open future.

We are left, with Gregory, to imagine what feelings may have flowed between Macrina and the iron ring. Did the metal feel cold or warm, light or heavy against her chest? we might wonder. Did it clink against the iron cross with which it was paired? Did contact with skin, sweat, and air cause the iron to corrode, or did it remain unchanged by such touchings? Did Macrina caress the ring with her fingers, kiss it with her lips? Did she open the receptacle to touch the fragment of wood directly? If so, to what desires of ring and fragment might she have been responding?[38] Did she talk to them? Did they answer? Who crafted the ring, and how did it come to Macrina? How many human relationships did it subsequently outlive? Where and how did it perish, or does it live still, whether slumbrous or vigilant? What, after all, is the life span of an iron thing with a heart of wood?[39]

Heart calls to heart, wood to wood. Each splinter of the "true cross" summons the whole, yet perhaps the parts are far greater than their sum. As early as the late 340s, some fifteen years after the discovery of the purported cross of Christ near Golgotha, Cyril of Jerusalem writes that "already the whole world is filled with fragments of the wood of the cross" (Catechesis 4.10). Through its fragmentation and dispersal, which humans aid like bees spreading pollen, the wood of life continues to grow—to reproduce, even. Some processes of fragmentation and dispersal are more violent than others: the late fourth-century pilgrim Egeria reports that when the cross was displayed on Good Friday in Jerusalem, one of the faithful "bit off a piece of the holy wood and stole it away" (Travels 37.2). Others stress not the cross's violability but its exceeding bounty. Writing in the early fifth century, Paulinus of Nola asserts that the cross has a "living vigor" in its "unsensing matter" and "has bestowed its wood upon the countless, almost daily, prayers of humans. Yet it suffers no diminution; though daily divided, it remains as if intact to those who lift it, and always whole to those who venerate it" (Letter 31.6). We note Paulinus's phrasing: it is the cross that has bestowed its wood, not its human handlers. Writing just slightly earlier, Victricius of Rouen makes the same argument for relics in general: the saints "do not inflict loss upon themselves by their own dissemination, but being endowed with unity, they distribute benefactions" (Praising the Saints 9). Indeed, the saints positively desire to fall to pieces so as to take themselves to new places: "The longing of the saints is not to be deferred!" exclaims Victricius. He has built a basilica for holy relics received from Ambrose of Milan, and the desire that has impelled such a building project arises as a response to the otherwise "hidden" longing of the saints themselves, as he understands it (12). In other words, his own desire reveals theirs. As his habit of referring to relics as "saints" suggests, Victricius insists that "the whole can be in the part," no matter how small, when it comes to saintly relics (10). Thus the capacity of the cross, or any other relic, to reproduce itself—or, in Victricius's terms, to reproduce the saint herself!—is virtually limitless. As Peers argues, "How one sees the regenerative potential in the world around oneself implicates all manner of thinking."[40] For Victricius and other late ancient Christians, matter was alive and fecund.

Inscribing himself into the circuitry of the cross's bounty, Paulinus sends his friend Sulpicius "a part of a particle of the wood of the divine cross" (Letter 31.1). This tiny sliver has traveled far, even before setting out from Paulinus's Campania for Sulpicius's Aquitaine. Victricius assures us that travel induces no weariness in the saints' relics (Praising the Saints 9), and perhaps that is due in part to the efforts of their human helpers. As Paulinus relates, Melania receives the fragment of the cross from John, bishop of Jerusalem, and subsequently carries it from Jerusalem to Italy. Whether she intends it as a gift to Paulinus or to his wife Therasia is unclear. Writing to Sulpicius, Paulinus claims the gift as his own even as he passes a bit of it on to Sulpicius. At the same time, he suggests that it is Therasia who sends it to Sulpicius's mother-in-law Bassula, harmonizing the two accounts as follows: "What is given to one of you belongs to the other, because one mind abides in both of you, and the faith by which you both meet 'unto a perfect man' [Eph. 4.13] eliminates sex" (Letter 31.1). Thus the fragment does not only travel great distances; in so doing, it also creates and solidifies relationships among humans. It unites Melania with Therasia and Therasia with Bassula; it unites Paulinus with Sulpicius; finally, in the conflation of both givers and recipients, it unites Paulinus with Therasia and Sulpicius with Bassula, thereby overcoming even the difference of sex, Paulinus stresses.

Most important, it unites all of these Christians with Christ, collapsing both temporal and spatial distance. Imagination and emotion transform the "scarcely divisible segment of a short little sliver" of wood into the very cross itself "on which the lord of majesty was nailed and hung while the world trembled" (Letter 31.1).[41] Jerome gives us an example of this upheaval of sense and feeling in his description of his friend Paula's encounter with the cross in Jerusalem: "She fell down and worshipped before the cross as if she could see the lord hanging on it" (Jerome, Letter 108).[42] For his part, Paulinus instructs Sulpicius on how to interact with the wood of the cross in order to achieve a similar experience. "You should look with inner vision"; "you should rejoice with trembling." Trembling, the viewer imitates the earth that quaked when Christ was crucified, suggests Paulinus. One should also "recall the rocks that split at the sight of the cross," he urges. "In imitation of the stones, let us split our hearts with fear of god" (Paulinus, Letter 31.1; cf. Matt. 27.51).[43] If the wood of the cross evokes the vision of Christ crucified, quaking earth and splitting rock teach how to respond to that vision. Here the biblical text offers a narrative world of nonhuman things, inviting humans to encounter them as exemplars. To see the crucifixion in the fragment of wood, one must become earth, become rock. One must tremble and fear. To see is to be transformed by emotion.

Like Macrina's, Sulpicius's fragment of wood is invisibly ensconced in a metal covering, in this case not humble, sturdy iron but fiery, fluid gold, at once malleable and resistant to corruption, as Pliny details (Natural History 33.19). "I sent you your own exemplar, in the figure of gold," Paulinus informs Sulpicius. He adds, "I know that you, like fired gold, have within you the kingdom of god, that is, the faith of the cross" (Letter 31.2). Fired gold harbors the kingdom of god and so does Sulpicius. The reliquary imitates Sulpicius, then, and Sulpicius himself imitates the reliquary. But what does that mean? A reliquary is not merely a container but a kind of *frame* that invites new ways of perceiving and relating to the thing it frames. As Peers argues, "framing passages define the objects as such,"[44] that is, as active, relational, and elusive (to evoke Heidegger's thingly triad yet again). The reliquary frames the fragment as an object or thing by revealing its shining beauty, and also by concealing it.[45] Page duBois proposes that "fragments suggest an aesthetics of distance, of the beauty of the unreachable object of desire, suggest the superiority of the unattainable."[46] The reliquary enhances such an aesthetics, ensuring that the sliver of wood cannot be directly seen, touched, or bitten but remains desirable in its very withdrawal. Yet the reliquary does more than distance and conceal: relic and reliquary together constitute an ever-renewing[47] assemblage in which the reliquary does not merely hide the relic but also renders it present, however indirectly, displaying and even amplifying its power and affectivity. Giving visual expression to "the call that arises from things,"[48] the reliquary invites and mediates relational engagement on the part of humans as well. When Sulpicius looks at the wood "with inner vision," he looks at the golden encasement, which signals the presence of the wood hidden within. With both inner and outer vision, then, he sees the fragment through the reliquary as a shining forth of what nonetheless eludes the grasp of both sense and intellect. In this seeing is a trembling and a sundering—a fragmenting—of the gazing subject as well. Fragment thus calls to fragment: in what Peers describes as an "absorptive process," distance and difference between the human looker and the thing that gazes back dissolves.[49]

Not all relics are encased in reliquaries, however. Gesture, for example, can also frame the relic as thing. As Miller notes, "Human body parts did not become the animate bodies that were relics apart from ritual practice."[50] The late fourth-century pilgrim Egeria recounts how the body of the wood of the cross discovered near Golgotha is displayed annually on Good Friday to the crowds that congregate: "They stoop down, touch the holy wood first with their forehead and then with their eyes, and then kiss it, but no one puts out his hand to touch it" (Travels 37.3). Some two hundred years later, another visitor describes the same ritual, now embellished with an account of the cross's powerful effects. "When the wood of the cross proceeds from its little room to be adored and comes into the atrium where it is adored, at that very hour a star appears in the sky and comes over the place where the cross rests," the Piacenza pilgrim reports. "And while the cross is adored it [the star] stays over it and oil is offered to be blessed in little flasks." The effect of these offerings is dramatic: "When the mouth of the flask touches the wood of the cross, the oil instantly bubbles over [*ebullescit . . . foris*], and if it is not closed very quickly it all overflows [*totum redundat foris*]." Once these acts of veneration have been performed, "the cross returns to its place" and so too does the star, explains the pilgrim, in closing his remarkable account (Piacenza Pilgrim 20).

Here again we encounter a dynamic assemblage of actors—the wood, which *proceeds* and *comes*, desiring adoration; the star, which also *comes*, as if summoned by the wood's desire; the flasks, which *touch* the wood in a caress of lips that causes the oil within them to *froth* and *spill*. The cosmic scope and affective, even erotic, intensity of this constellation is striking. So too is the fact that humans seem to play merely supporting roles, and even those are not directly named. Although the pilgrim initially identifies the wood of the cross as that "which[51] we adored and kissed [*quod adoravimus et osculavimus*]," human agents quickly disappear from the account. It is the star who arrives in response to the cross's desire to be adored and the flasks who offer their kisses and ejaculatory outpourings.

One might counter that the actions of humans are everywhere implied. What, then, are those actions? The cross desires to be adored, and the humans respond, bringing it out of its little room and uncovering its nakedness, exposing it to the heavens, touching it with their eyes and their lips. Or does the human desire to adore come first, awakening the desire of the cross to be adored? Such a question cannot be answered and not only because of the inherent elusiveness of the feelings or wants of things: whether human or nonhuman, desire ever responds to desire, and it is impossible to say where it starts. But the human actors do not only engage the cross directly. They also participate in the adoration of the cross performed by the flasks, enabling their mobility, marveling at their expressive ebullience, cherishing the holy power transferred from the divine body of Christ to the receptive media first of wood and then of oil.

How are we to imagine these modest yet lively containers in their material specificity? At the Israel Museum in Jerusalem, a placard proposes that our pilgrim's *ampullae* might have resembled a pair of pottery "juglets" on display, roughly three inches high, with open lips, long necks, and swelling bodies that taper almost to a point (Figure 6).[52] Such flasks would fit comfortably in a human hand. Moreover, their shape, together with their viscous, overflowing contents (as the Piacenza pilgrim imagines them), renders them ambiguously erotic. That semen is a frothy substance is a long-standing tradition going back to Aristotle; intriguingly, the philosopher compares the foam of semen, understood to be a residue of blood (On Generation 726b 1–12), to the foam produced by the vigorous mixing of oil, water, and *pneuma*, or air (735b 13–16, 736a 16–18). At the same time, the physician Soranus describes the uterus as possessing a bottom, sides, shoulders, neck, and mouth (Gynecology 1.9), an anthropomorphism also applied to jugs or flasks; indeed, ancient sources sometimes compare the uterus to an upside-down jar.[53] As a container, it both gives and receives, excreting fluids while also receiving semen through its hungry mouth (Gynecology 1.10). The point is not that the earthy flasks are either phallic or uterine but that they potentially evoke an unstably gendered array of erotic associations.

But perhaps the *ampullae* described by the Piacenza pilgrim were not made of clay but rather of a hardier material like metal. Consider a lead flask

Figure 6. "Eulogia Juglets." Israel Museum, Jerusalem. Photographer, Max Peers.

from late sixth- or early seventh-century Palestine, roughly contemporaneous with our pilgrim.[54] Unlike the unadorned pottery vessels, this object shows and tells quite forthrightly. As Miller puts it, it has an explicit "epistemological function."[55] One side of the flask's round, flat body depicts a crucified Christ flanked by the two thieves believed to have been crucified with him; two worshipers kneel at his feet. Circling the image are the words "oil of the wood of life from the holy places of Christ." The other side depicts the ascension of Christ, flanked here not by thieves but by angels; Mary and the apostles look on. This second scene is glossed with the words "blessing of the lord from the holy places" (Figures 7a and 7b). The *ampulla* thus announces its otherwise invisible contents—oil that has come in contact with the wood of the cross, carried away from the "holy places" of Jerusalem as a gift or "blessing." It also evokes the events that imbue both the oil and the places with sacred power—Christ's passion, resurrection, and ascension. Finally, it inserts the pilgrim into these scenes. As Derek Krueger notes, "The kneeling figures and the watchful Mary and the apostles provide models for the pilgrim's own religious response."[56] Commenting on a similar *ampulla*

Figure 7. "Pilgrim's Ampulla with Scenes of the Crucifixion." Palestine, early Byzantine period, c. 600. Tin-lead alloy with leather fragments; overall, 6.3 × 4.6 × 1.5 cm ($2\frac{7}{16} \times 1\frac{3}{4} \times \frac{9}{16}$ inches). Courtesy of the Cleveland Museum of Art, John L. Severance Fund 1999.46.a–b.

featuring a crucifixion with two kneeling figures, Peers likewise suggests, "Not part of the historical narrative, they [the kneelers] instead represent the assimilation of viewer and object of devotion into a permanently commemorated relationship."[57] Thus the flask not only facilitates the oil's mobility, thereby further disseminating the power of the wood of the cross; it also mediates the pilgrim's relationship to the holy oil and, beyond that, to the cross and to Christ. Like a reliquary, the flask simultaneously hides the oil and renders it potently present. Lying in the palm of a pilgrim's hand or hanging from a leather cord around her neck, it invites the intimate touch of fingers and eyes. Its expressive surfaces speak to the pilgrim through images and letters, telling of oil, wood, cross, and the crucified savior who heals and enlivens all things.

Late ancient Christian texts and artifacts thus give witness to what Miller has aptly dubbed a "material turn" in Christianity.[58] As we have seen, this turn is nowhere more evident than in the proliferation of mobile, tactile *things* imbued with holy power—relics and images of Christ and his saints. By cultivating special relationships with such supercharged objects, Christians likewise cultivate particular sensibilities regarding material things more generally: this simple point is at the heart of my argument.[59] In the realm of Christian practice, relationships with nonhuman things appear no less important than relationships with other humans; moreover, the former are intimately interwoven with the latter. At the same time, relationships to holy things both intensify and ultimately undercut the distinctions between the material and the spiritual, the natural and the artificial. Bone, wood, blood, oil, wax, metal, glass: ancient Christians participate in "a world of active materials."[60] In the making of holy things, human craft colludes with the ongoing flow of divine creativity, as the vital power of materiality traverses the boundaries between human and nonhuman, creature and god.

The relations of humans to relics and images is played out above all in the realm of touching and feeling. Following Aristotle, Jean-Louis Chrétien proposes that touch "is the most fundamental and universal of all the senses," being shared by all animate beings; indeed, it is "not primarily and perhaps not even ultimately one of the five senses" but rather is basic to sensation itself. Intensely interactive, "the sense of touch, far from making the living organism into a mere spectator, pledges it to the world through and through, exposes it to the world and protects it from it."[61] But what counts as an animate being or a "living organism"? we might ask. For ancient Christians, anything *might* count. Iron may not see or hear, but in the form of a ring on a necklace it touches Macrina's skin and her skin touches it. Such intimate proximity seduces with the illusion of immediacy, even as it also inscribes distance. Touch does not, after all, break down the difference between touching things: Macrina remains Macrina, ring remains ring.[62] Paradoxically, however, what may appear to be even more obvious and extensive mediations of touch yield a queer sense of immediacy where desire strains to close the gap. Skin touches ring, ring touches splinter, splinter touches cross, cross touches the body of Christ: such distancing intervals notwithstanding, however intimate the touching of skin and ring may feel, the touching of Macrina and her divine bridegroom is the most intimate—and the least mediated—of all.

Such touching may also be visual, moreover: as Peers notes, it is difficult for many of us "to understand with full implications . . . the haptic quality of seeing" as ancient folk experienced it.[63] When Sulpicius's eyes rest on the golden reliquary encasing a tiny fragment of the wood of the cross, they also brush the body of the crucified Christ hanging on that cross, as Paulinus urges. Touching with his eyes, Sulpicius will be touched as well, dazzled by the gleaming brilliance of humble wood and tortured flesh, quaking and shattering like earth and stone at the sight. This kind of seeing is no mere spectatorship, in other words. It is an interactive process of sensation that opens onto a depth of feeling. *Any* thing and *every* thing may potentially draw near, even unbearably near. Hierarchies of being, including the distinction between living and nonliving, do not hold. Nor do hierarchies of beauty hold. *Any* thing and *every* thing may potentially dazzle the senses, overwhelm the emotions.

By cultivating sensual connections with relics and icons, late ancient Christians thus open themselves to a wider world of lively, tactile, and irreducibly relational *things*. These *things* are experienced as neither static nor self-contained but rather as mobile, fragmentary, and ever-emergent. In place of stable orders, hybrid assemblages coalesce and dissolve. There a nonhuman thing may be encountered as a supplement for a human, while a human may equally be encountered as a supplement for a nonhuman. In Peers's evocative phrasing, "We are objects' prostheses, their way to overcome their physical limitations and to realize their own emotional, sensory lives more fully than they can on their own."[64] Humans may become the vehicles for the movements of relics or icons; we may clean or adorn or repair their bodies; we may show them love through caress of word or touch. They in turn light up our senses and emotions, embodying the vitality of materiality and the openness of things to one another.

Fragments par excellence, relics and images puncture the false closure of an imagined wholeness. They do so first by paradoxically claiming wholeness for every part, no matter how infinitely small or abundantly multiplied, and second by imbuing the part (now torn from any predetermined totality) with both deep memory of and intense longing for the other. But which other? Any other? The fragment "intervenes as a perpetual opening to alterity," as Leslie Hill puts it.[65] Haunted by what is other and more than themselves, relics and icons reveal the relational dynamism of a fragmented world of becoming, in which every part, so to speak, both withdraws from and longs for every other part (albeit not identically). Channeling the desires of the humans who cherish them, these Christian objects open themselves to otherwise imperceptible presences—the presences of the saints, of Christ, of the divine itself. In so doing, they open their humans to the reality of "a collaborative, relational materiality"[66] at once vibrant and mysterious. Ring, cross, iron, wood, blood, gold, earth, stone, bone, star, flask, sky, oil, eye, skin—an intimate, alien swarm of feeling, felt *things*.

Situating Things: Architecture, Landscape, Cosmos

We have been focusing on Christian things in their most tactile mobility—things carried or caressed, for example. Some of those things are already on their way to joining larger, more stable assemblages. We recall, for example, the fragment of wood that travels from Jerusalem to Aquitaine, destined to be housed in the altar of Sulpicius's newly constructed basilica (Paulinus, Letter 31.1). The relic's relative loss of mobility entails some sacrifice, and Paulinus acknowledges the human cost in a letter to his friend: "You may desire to have this blessing from the cross available for your daily protection and healing, and once it is buried within the altar it may not be always accessible according to need" (Letter 32.8). For the relic too, there is a corresponding diminishment of intimate, immediate connection. What, then, are the gains for relic and human alike that might offset such poignant losses? Put simply, by taking up residence in the basilica, the fragment of wood becomes part of something bigger, in both scale and complexity; its relations are multiplied, its agency amplified. Experiencing the relic in such an enhanced and enhancing setting, Sulpicius himself will enjoy new relational benefits, not least with the basilica he has built on his own estate and potentially with the world beyond it as well.

Like any building, a church is a mesh of things and their relations—relic, altar, and so much more—yet it is also a thing with its own relations, "exert[ing] a force on the world independent of human intention or even human consciousness," as Annabel Wharton puts it.[67] As C. M. Chin points out, a building typically enjoys a longevity that far exceeds that of the humans who have "built, maintained, and passed it on to others." Its extensive life span demands "the protection of new caretakers," thereby pressing human behavior into accord "with nonhuman duration."[68] A church enjoys significant relations not only with the humans who move through it and care for it, however fleetingly, but also with the environment, both built and natural, that surrounds it, more enduringly. Yet conventional concepts here partly fail us. As with the smaller, more volatile things we have considered, the line between human and divine creations—what I have just called the built and the natural—begins to dissolve on closer inspection. At the same time, *environment* begs to be understood not as the surroundings of a particular subject or object but rather as the long, slow dance of mutual encircling in which all things participate. A cosmic liturgy, perhaps?[69]

A column found at Tel-Dor on the Mediterranean coast of Israel some six miles north of Caesarea offers material witness as to how a Christian holy thing might come to be incorporated into a larger collectivity. Currently standing on display at the Mizgaga Museum in Kibbutz Nahsholim, the roughly human-sized pillar addresses the viewer in its own engraved words (Figure 8). Inscriptions that animate the surfaces of late ancient buildings and other objects endow those surfaces with a kind of speech, as Amy Papalexandrou has argued, "sometimes addressing the reader directly, sometimes referring back to themselves." The inscribed object speaks most audibly in the borrowed voice of the one who reads the inscription aloud—standard practice in the oral cultures of Mediterranean late antiquity, as Papalexandrou notes[70] (or, not dissimilarly, in the context of an out-of-the-way museum where a visitor might find herself puzzling through a bit of ancient Greek). Yet perhaps the inscribing hand embodies a kind of voice as well. As Jean-Luc Nancy puts it, "Even when written, [sense] has a voice."[71] If so, our column seems to speak in declarative tones more familiar then formal. Clearly but roughly drawn, the carved letters drift slightly out of alignment as they follow the curve of the gray marble surface, and the three-line message is brief and to the point: "A stone of holy Golgotha" (Figure 9).

† ΤΟΥΑΓΙΟΥ †	† OF HOLY †
ΓΟΛΓΟΘΑ	GOLGOTHA
ΛΙΘΟϹ	STONE

Two crosses framing the first line point to the wood of the cross in Jerusalem, an association further secured by the reference in the second line to Golgotha; the third line tightens the reference to a stone of Golgotha. One might wonder whether the column is naming its own identity or at least referencing its own genus: it is, after all, made of rock. That rock did not likely come from Golgotha, yet when a stone column in Dor speaks of a stony hill in Jerusalem where Christ is said to have been crucified, is a certain lithic affinity not implied?

As the reading eye is tugged by its own downward momentum, it encounters more crosses. The largest is discernible only as a ghostly absence hailed by the inscription. At its center is a gouge in the marble that apparently once held a small reliquary box, riveted in place by a cruciform plaque; the latter

Figures 8 and 9. "Reliquary Column." Inscribed gray marble column, early Byzantine, Tel-Dor. Mizgaga Museum, Israel. Photographer, Max Peers.

has left its imprint on the column but is now lost, as is the reliquary itself, along with the fragment of Golgotha stone that it likely housed (Figure 9). This particular reliquary cover may have replaced an earlier one, as it seems to have partly obscured four smaller crosses matching the ones in the inscription, carved at each of the ends of the central cross. Whatever their exact history, such a layered abundance of cruciform figures practically shouts: it is the cross of Christ that frames the significance of the stone announced in the inscription. If Golgotha is the place where the cross was once planted, its rock is infused with the memory and power of the wood of life, and the memory of rock runs deep and long. A chunk of that living rock is in turn embedded in a polished column that stands as straight as a cross. Nature and artifice are thereby conjoined, and so are this place and that: there is a bit of Golgotha in Dor.[72] Is it also significant that the plan of the fourth-century basilica of Dor recalls the grander Constantinian basilica built on the site of Christ's passion and resurrection in Jerusulam?[73] Localities are entangled with one another in mysterious ways, both column and church suggest. They may even be rooted in the same bedrock. After all, Golgotha was understood to be the site not only of the crucifixion but also of creation, a Christian *omphalos* of sorts: from its foundational stone flow life-giving waters.[74]

Lithic time moves much more slowly than human time, and rocks riveted to marble columns might be thought to have achieved a high degree of stability, but no thing endures forever. The living rock of Golgotha and its reliquary have disappeared from the ruins of Dor, and the column itself was no longer standing when it was discovered in the mid-twentieth century; indeed, it appears to have attempted to flee the scene as well. Joseph Leibovitch reports as follows regarding his 1952 excavations of Tel-Dor: "While inspecting the surroundings of the spot, I was lucky to find, about a hundred metres east of the basilica, a gray marble column lying on the floor. This column did not seem to be *in situ*. It had neither capital nor base."[75] Archaeologists have subsequently speculated that the column may once have stood in a small apsed chapel located in an external aisle on the south side of the basilica—perhaps an echo of the positioning of Golgotha within the Jerusalem basilica complex, then. The chapel abuts, and may have opened onto, a tomb in the interior aisle where two human bodies are buried. One of the slabs covering the tomb is pierced by a small hole lined with a pipe through which oil would have been poured, draining into a basin at the bottom of the tomb, where it could be collected to be distributed as a blessing and used to heal the sick.[76] Thus the column would have found itself in good company with oil-dispensing saints as neighbors, and the basilica would surely have attracted many Jerusalem-bound pilgrims arriving either by sea or along the Via Maris, or coastal road.[77]

Perhaps the column was part of another building in a previous life: the basilica site sits atop remains of an early temple complex thought to have been dedicated first to Apollo and later to Asclepius, and stone slabs from this structure were reused in the building of the church.[78] In its Christian location, would it have carried the weight of an arch, with its graffito-like inscription and reliquary adornment a sort of surplus spilling beyond its architectural role?[79] Or would it, alternately, have been encountered as a free-standing object, strategically positioned so that pilgrims could circulate past it, reading its words, touching its smooth marble skin, bending to kiss its (navel-high) relic? Would it have glowed warmly in sunlight from a window, danced playfully in flickering flames of candles, shrouded itself seductively in shadows? Wherever it was situated and however it presented itself to the eye, it would have played its role in the sensuous flow of the church, drawing visitors with the magnetic appeal of its rough, precious stone, providing a pivot point for their circumambulations, releasing them to the next object of attraction—perhaps the saints' tomb or the large peristyle court and porticoes that are thought to have sheltered the sick who came to the basilica for healing in the time-honored practice of incubation.[80] Our column would, in other words, have performed as an animate being participating in the life of a larger animate being, the basilica of Dor.

Relatively little survives of that seaside basilica, on which new structures and lives have encroached, just as the basilica once encroached on others. (In fact, the pagan temple on the site appears to have been burned and razed before the construction of the Dor basilica.)[81] Most of the remains of the southern external aisle, where the column's chapel is thought to have been located, were destroyed in the mid-twentieth century by the construction of houses in Kibbutz Nahsholim.[82] We must, then, turn elsewhere to imagine more fully the life of a late ancient basilica adorned with inscriptions and images, thronged with human visitors, configuring and configured by relations both interior and exterior. Here the writings of the wealthy senator Paulinus are especially illumining. In the early years of the fifth century, Paulinus undertook both renovations and new construction at the shrine of Saint Felix near Nola, and his letters and poems offer a rich view of the relational life of the so-called *basilica nova* in particular.

Paulinus has a fine sense of the liveliness of things large and small, as we are reminded in the very first line of a letter addressed to his friend Sulpicius in 403 or 404. He has just sent Sulpicius a letter accompanying the gift of a fragment of the wood of the cross; however, he is already impelled to write more, he confesses. Why? Because the blank parchment seduces him! "The open page [*patens pagina*] made advances to [*sollicitavit*] my tongue and hand to fill out the empty spaces," he declares. Temptingly exposing itself, the receptive paper provokes his outpourings, and Paulinus is moved to celebrate his intimacy with Sulpicius. This intimacy results in part from their parallel endeavors, which make them as if one in heart and body: Sulpicius too is a constructor of ecclesiastical buildings (Letter 32.1). Indeed, the basilicas of these two builders are alter egos of a sort: like Paulinus and Sulpicius, they long to be together. Paulinus tells Sulpicius that he is moved "to unite [*compaginare*] my basilicas with yours in writing." Their buildings will be on the same page, then, "just as they are already conjoined by the timing of their construction and the manner of their dedication"; although separated by great distance, they will be brought close "by a chain of letters" (32.10). If the basilicas need Paulinus's letters to unite them, Paulinus seems to need the basilicas to draw himself nearer to his far-away friend through the same chain of letters. Making use of scriptural imagery, he imagines that he and Sulpicius are themselves buildings: "For even as we erect these buildings in the Lord because we have received the faith, we are ourselves erected by the Lord through the growth of this same faith" (32.18). Similarly, "So let us entreat the Lord that, while we build a visible lodging for him outside ourselves, he may build within us a lodging which is invisible" (Letter 32.23; cf. 2 Cor. 5:1). Metaphorical thinking? If so, not *mere* metaphor.[83] Paulinus is able not only to think with buildings but to think as a building, and to imagine that buildings think—and desire.

His own basilica not only thinks and desires but also speaks. Built near the tomb of the confessor Felix and dedicated to the apostles whose relics it contains, it has plenty to talk about, as we shall see. Paulinus initially directs attention to the tripartite apse, a natural focal point. "A vault adorned with a mosaic brightens the apse," he notes. However, rather than describing the mosaic itself, he cites the poetic verses that he has written to accompany its *picturae*, verses that are apparently inscribed in the vault. Thus the vault not only lights up the apse but also simultaneously shows and tells. Its telling is itself a kind of showing, moreover. "*Pleno coruscat trinitas mysterio*," the poem proclaims: "The trinity glitters in its full mystery!" The shining of the vault mosaic is thus conflated with the flickering brilliance of the triune god. No other source of light is mentioned; thus the reader is called upon to imagine not a passively reflective surface but an actively radiant one. (In fact, the archaeological remains suggest that the apse was perforated by windows, but the verses ignore this feature.)[84] The mosaic glows, then, which is also to say: the divinity glows. Now *picturae* begin to form within the dazzle: "Christ stands [*stat*] as lamb; the voice of the Father thunders [*tonat*] from heaven; and the Holy Spirit glides [*fluit*] in the form of a dove." The first and third images in this triad can be easily visualized—lamb and dove, the one firmly planted, the other moving and fluid. The second image is not seen but heard—a thundering voice, descending from above. A biblically literate reader or viewer will recognize an evocation of the scene of Christ's baptism, with its heavenly voice and dove (cf. Matt. 3:16–17; Mark 1:10–11; Luke 3:21–22), though the lamb seems to have wandered in from a slightly different set (cf. John 1:29). Another image swiftly joins this triad—a luminous chaplet encircling a cross, itself encircled by apostles appearing as doves. Thus the figure of Christ is doubled in lamb and cross, representing his victimhood, while the doubled signs of purple and palm mark his triumph, we are told. Indeed, the figure of Christ is not just doubled but tripled, it would seem, for he is also a rock, the verses announce (cf. 1 Cor. 10:4)—a rock standing upon the rock of the church (cf. Matt. 16:18, but also Eph. 2:20), from which issue four noisy fountains (cf. Gen. 2:10), identified with the evangelists, "living streams of Christ" (cf. John 7:38). The pattern of standing, resounding, and flowing thus repeats itself (Paulinus, Letter 32.10).

It is possible to map these poetic images onto a two-dimensional visual plane with a defined artistic and doctrinal program, thereby gaining some control over their appearance and meaning, should that be our goal. Several sixth-century Italian *comparanda* aid in such an endeavor. Consider Rudolf Goldschmidt's description: "In the Nolan mosaic we imagine a cross with a wreath of doves around it, the Father's hand coming down from Heaven from between the clouds . . . and immediately under this above the Cross the Dove. Palm trees are visible in the landscape. A purple cloth lies on the Cross." The dominant elements depicted here appear in the apse mosaic of Sant' Apollinare in Classe, the central figure of which is a gold cross encircled by a jeweled wreath. Above the wreath descends a hand from a gold sky striated with pastel clouds, offering visual representation of the divine voice, while below the wreath stretches a gardenlike landscape traversed by rows of trees.[85] Missing from Sant' Apollinare are only the doves and purple cloth, but for the purposes of imaginative reconstruction, these may be imported from elsewhere: the Albenga baptistry in northern Italy depicts a circular chi-rho surrounded by twelve doves,[86] while the Arian baptistry in Ravenna includes both a cross draped in a purple cloth and a dove that "flows" (cf. Letter 32.10)—gushing either water, oil, breath, or light as it descends from the heavens.[87] Goldschmidt continues, "Under it [the wreathed cross] the Lamb is standing on a rock, from which issue four rivers: on either side may have been six sheep, not mentioned in the inscription, but belonging to the scheme."[88] For these elements, we may turn to the lower register of the apse mosaic from Santi Cosma e Damiano in Rome, which depicts precisely such a scene, set against a gold background.[89] Cutting and pasting thus produces a plausible design that correlates with the poem's visual cues, and also with its theological ones. Hand and dove at the top, lamb at the bottom, all are set off by gold tiles, according to this reconstruction: in these images, "the trinity glitters." For its part, the centrally positioned, wreathed cross is where "the holy unity of the trinity meets in Christ." Baptismal themes announce Christ's divinity, "revealed as god by the paternal voice and the spirit," while the figures of cross and lamb highlight the sacrifice performed by his humanity, purple cloth celebrating his ultimate victory. Paradisal themes point toward the salvation made available through Christ's church, represented by the apostolic "chorus of doves" and the foundational "rock" and "living streams" of the evangelists (Letter 32.10).

What here claims our primary attention, however, is the extent to which the *ekphrasis* recited by Paulinus exceeds and evades such a flat mapping of the visual and theological register of the mosaic through its lively performativity. Who speaks? Not Paulinus the letter writer. Paulinus the poet, then? Perhaps. But the poetic voice, distanced from that of the letter writer, is thereby also merged with that of the basilica itself. The vaulted apse proclaims in glittering images; they enter through the eye but also tickle the ear—a point that our cut-and-paste visual map misses. The paternal voice thunders, the fountains splash. The poem itself sings. It sings *picturae*, biblically saturated—and exclusively nonhuman!—figures, released into a mobile, multidimensional space of sensation and feeling. Lamb, voice, dove, cross, wreath, purple, palm, rock, rivers: all of these exert their force. Coiled with the energy of its active, present-tense verbs, the mosaic-poem finally impresses less by its appearance or design than by its sheer affectivity. Is the apse vault using its words to teach human spectators how to be moved by its own nonverbal speech? Or is it simply pulling out all the stops, piling (one kind of) image-speech onto (another kind of) image-speech as it clamors for attention? A nonliterate viewer would not be able to give voice to the verses, but the mere presence of the inscribed letters might still effectively perform a kind of talking. Whether literate or not, the viewer is not so much interpreting the mosaic and its inscription as being interpellated by it. *Hey, you!* it cries. (Of course, she may always choose to resist the call.)

Indeed, the basilica hums with speech. Other parts speak in voices more straightforwardly informative or, alternately, more explicitly imperative: Paulinus's building can be bossy! As a visitor draws close to the apse, she is addressed by words inscribed on the border between vault and wall, directing attention to the relics under the altar. "Here is the cross, joined with its witnesses [*martyribus*]," the church exults. Here it is, but it hides from the gaze. "*Velant altaria*": "the altar covers up" the wood of the cross, as well as the apostles' ashes and bones with which it mingles. Nonetheless, the viewer is invited to imagine both the smallness and the greatness of the "little splinter" that is at the same time a "big pledge," promising to deliver "in its tiny bit the whole power of the cross" (Letter 32.11). The fragment has a particular history and pedigree, and these are also here announced: "Brought to Nola by the gift of holy Melania, this greatest blessing came from the city of Jerusalem." It has relationships as well, not only with Melania who carried it but also with the apostolic martyrs with whose remains it now rests, with the altar that houses it, the verses that name it, the visitors who come to adore it, the mosaic above that depicts the cross, the cross itself in Jerusalem, and all the other dispersed fragments of that cross, not least the sliver sent by Paulinus to Sulpicius—to name only the closest and most obvious relational possibilities.[90]

Apse and altar are, then, particularly active and articulate sites in Paulinus's basilica, as we might expect. So too are thresholds. Doorways greet: "Peace be with you, whoever enters the inner sanctuary [*penetralia*] of Christ the god." They bid farewell: "Whoever leaves the house of the lord, after completing your prayers in due order, depart in body, remain in heart." They tell visitors what to anticipate and what to remember in crossing from one place to another. Opening onto the church from "a little garden or orchard," a private gate declares, "Approach heavenly paths through pleasant lawns, Christ-worshippers [*christicolae*]! An entry hence from joyful gardens is also fitting, whence an exit into holy paradise is given to those who deserve it" (Letter 32.12). Thus the church itself becomes a portal, leading from worldly to heavenly landscape.

Doorways are also the basilica's eyes, communing with its exterior. The "view" (*prospectus*) of the basilica is to the south, "looking toward [*adspiciens*] the memorial of Felix" through its "threefold door." Indeed, Paulinus has gone to great trouble to open up this line of sight, from basilical apse through nave and doors to tomb. Previously the saint's tomb was enclosed, but the obstructing wall of the old church has now been laid open (*patefactus*) by three doors aligning with those of the new basilica's entrance. The dual sets of triple doorways form a "translucent lattice" (*perlucens transenna*), which "offers a vista, filigreed as it were [*quasi diatritam speciem*], from one church into the other" (Letter 32.13). "Within the lattice"—that is, in the interval separating the two visually connected buildings—flows a "glistening fountain [*cantharus*]."[91] Across this lacy, sun-dappled courtyard, the churches, old and new, gaze at one another: "Through tripled arches twin halls lie open, and they admire [*miranturque*] each other's garb [*cultus*] across shared thresholds" (Letter 32.15). And why shouldn't they? As Paulinus notes elsewhere, even the *basilica antiqua* shines with refreshed beauty, having donned "new garments purged of age" through Paulinus's own renovations (Hymn 27.383).

Paulinus's *basilica nova* thus not only speaks but also looks. It looks lovingly at the tomb of Felix through the latticework of archways; it looks admiringly at the older church that houses the tomb. We may assume that it also looks at the courtyard that lies between them. And what does it see there? A friendly face, it would seem: in Hymn 28, Paulinus asserts that the courtyard "smiles [*ridet*] with its varied decorations." He notes again the shining central fountain (*cantharus*), as well as myriad "vases," "little fountains," and "basins," all flowing with water (Hymn 28.28–52).[92] In Hymn 27, he similarly (if also more succinctly) describes the courtyard as "filled with so many fountains [*fontibus*]," here specifying, with attention to technical detail, that they are all fed by cisterns. In water-deprived Nola, the local waterworks rely on "fountains [*fontibus*] from heaven"—god-given rainwater, that is—rather than the "earthly aid" of aqueducts, he explains; sometimes these fountains flow, and sometimes they do not (Hymn 27.463–76). Thus the courtyard does more than mediate the relationship of the two churches. It also relates the basilica to the cultivated and adorned landscape, and thence to divine creation writ large. It does so along a specific visual axis, moreover: the line of sight running from apse to tomb links one fountain-endowed landscape to another.

As we recall, the apse mosaic features a rock "from which four noisy fountains [*fontes*] issue" (Letter 32.10). The apse itself looks out onto the landscape of "fountains [*fontes*] flowing in pleasant rivalry with gentle murmur," as Paulinus describes the courtyard (Hymn 28.48–49). If the apse looks out from within, a visitor may be imagined to follow a reverse path. Proceeding from outside in, she travels from "joyful gardens" to the "holy paradise" embodied by the glowing mosaic (cf. Letter 32.12). Although the basilica may seem to turn inward and away from the outside world, the inside also folds back onto the outside. At the same time, nature and art converge: *fons* can designate a spring—as in the mosaic—as well as a human-made fountain—as in the courtyard. An arch in the courtyard, pointing in the direction of the new basilica, proclaims, "To stunned eyes a new light opens, and the one who stands on the threshold beholds twin halls at once" (Letter 32.15). We may imagine eyes dazed by the sun of the courtyard, as well as the doubled riches of the coupled churches. And to be dazed by the sun of the courtyard is also already to be dazed by the glitter of the mosaic inside Paulinus's basilica. There world and paradise converge, and the line between interior and exterior, thus also between human and divine creations, blurs. As Gaston Bachelard puts it (with reference not to Paulinus but to the poet Georges Spyridaka): "A house that is as dynamic as this allows the poet to inhabit the universe. Or, to put it differently, the universe comes to inhabit this house."[93]

The question of how images of nature perform and interact in the context of late ancient church buildings deserves further exploration. Only small fragments of the mosaics and other adornments of Paulinus's basilica survive, so we must look elsewhere if we want to engage material images directly and in context. Santa Costanza of Rome offers a promising starting point, both because it is one of the earliest instances of monumental Christian architecture and art and because its structure and vault mosaics are particularly well preserved.[94] Likely built in the 340s as an imperial mausoleum in which two of Constantine's daughters would be buried, the domed rotunda was, like Paulinus's church, part of a larger complex that in this case included a large ambulatory basilica dedicated to the virgin martyr Agnes, constructed alongside the tomb of Saint Agnes herself.[95] Situated on a green slope descending from the Via Nomentana just outside the ancient city walls, the rounded shapes of both mausoleum and basilica appear almost to be organic extensions of the hill. This is an effect no doubt accentuated by the current state of the basilica: firmly rooted in the hill by massive buttresses up to eleven meters high in some places, only a portion of its brick and tufa walls remain, their lines softened by time. The dome of the contrastingly intact mausoleum now rises above its neighbor, which would once have dominated the landscape. And while one would originally have entered the mausoleum through the side aisle of the basilica, one now enters it directly from an open-air courtyard.

Regardless of how one enters, Santa Costanza creates a strong and immediate sense of enclosure. The bright interior space, illumined by the ample windows of the central dome, is set apart by the shadowy ambulatory that encircles it, lit only by the narrow slits of the clerestory windows and any lamps or candles that might be burning. Twelve pairs of columns support an equal number of arches that separate the ambulatory from the interior. That these columns are repurposed—thus, bring with them previous histories of life in other buildings—may, curiously, only enhance the sense of timelessness that seems to pervade the space. While some visitors will be drawn to the light and spacious height of the dome (whose mosaic unfortunately has not survived), others will find themselves immediately pulled into the circuit of the ambulatory, not least owing to the mesmerizing beauty of the mosaics that cover its vault, drawing the eye upward and forward. Were this building as talkative as Paulinus's basilica, it might say (in so many words), "Now I want you . . . to tire yourself [*fatiges*] a bit with neck thrown back, while you survey everything with face upturned" (Paulinus, Hymn 27.512–13). With or without such verbal cues, the decorated surfaces exert their force on the human bodies that view them, eliciting wonder at the expense of mild physical distortion or distress. High, arcing, and wrapping, the vault mosaics cannot be taken in at a single glance; they both entice and place strong demands on their spectators.[96]

It does not matter whether one turns left or right, for the mosaic programs are identical and symmetrical, progressing from the entrance to the rectangular niche on the opposite side, where the empress Constantina's porphyry sarcophagus likely once stood. Over the entrance is a field covered in a tightly geometric pattern of crosses with arms of equal length, alternating with octagonals containing floral chaplets encircling four-pointed flowers. Christian themes may (but need not) be inferred, even as the structure of the cruciform rotunda is echoed in the play of crosses and circles. On either side of the entrance are fields of diamonds and four-point stars, more intricate and less static in their rendering. There follow mosaic fields segmented by looping scrolls: large medallions contain anthropomorphic figures of winged genies and cupids; smaller ones, four-pointed plant or floral figures; and concave octagonals, birds and other animal figures. The proximate fields, covering the transversal axis, are also explicitly figural, while largely released from the control of geometric patterns and partitions: grape-picking cupids and birds clamber and perch among the curling tendrils of the vines that cover most of the surface; a human bust peers out from the center (male on one side of the mausoleum, female on the other), while the two bottom edges of each mosaic depict identical scenes of the transport and treading of grapes. Would a biblically literate Christian viewer see Christ and his followers in the sprawling vine? "I am the vine; you are the branches" (John 15:5). Might the Dionysiac scenes of winemaking also evoke eucharistic themes? "This is my blood of the covenant, which is poured out for many" (Mark 14:24). The next fields are

Figure 10. "Branches, Birds, Vessels." Ambulatory vault mosaic, fourth century, Church of Santa Costanza, Rome. © Vanni Archive/Art Resource, New York.

once again more tightly geometrical, being covered with identical medallions in which floral figures alternate with human figures and busts. Finally, the fields flanking the main niche are the most elaborate and striking in both form and color. Contrasting with the sense of containment created by the tidy rows of circular frames in the previous segment, here a variety of gold-flecked branches, birds, and luxurious vessels, horns, shells, and mirrors are strewn with artful abandon across an undivided background; the far edge of the mosaic field, abutting the light well in front of the main niche, is bordered by a glittering gold band adorned with images of inset jewels (Figure 10).

Expert opinion designates the motifs of the vault mosaics in the ambulatory of Santa Costanza as non-Christian, or at least not distinctly Christian,[97] yet I find myself sympathetic to the suggestion of environmental studies scholar Susan Power Bratton that "the vineyards and gardens [that] originally formed a circle around the biblical subjects of the central dome" might imply "a conscious Christian program," as the explicitly Christian narratives once depicted on the central dome are placed "within a great wheel of earthbound images, which are simultaneously symbolic of eternal life."[98] A viewer walking around the ambulatory would not only experience the circling of the "great wheel" overhead but might also discover in its shifting fields a series of juxtapositional translations. From the relatively static balance of cross and circle, death and life; to the more dynamic starbursts; to the array of *things* (animal, vegetable, human, divine), at once discrete and interconnected; to the curling, twining vines offering the sensual pleasures of fruit and wine—an arc of intensifying liveliness and complexity is traversed.[99] Perhaps not insignificantly, the most static field is the one that introduces human figures within the ordered containment of circular frames, as if raising (but not answering) the question of their interaction with one another, as well as with the more vivid mosaics that flank them—the vines with their bucolic scenes, on the one hand, and the exploding abundance of glistening, uncontained things, on the other.[100] The latter fields surround a light well that overlooks and illumines the space in front of the shadowy, star-studded niche that holds a wine-colored and vine-covered sarcophagus, thereby leading the viewer toward a culminating convergence of death and life, darkness and light, earth and spirit, nature and art, and nonhuman and human.

Perhaps, in the end, the vault mosaics of Santa Costanza are less "earthbound images" that are *also* "symbolic of eternal life," as Bratton has it, than images that communicate the excessiveness or surplus of vitality that overflows all sensuous things. As such, the mosaics walk their awkwardly crane-necked visitors through a process of coming to see the world more clearly, in all of its vibrant intensity and poignant transience—animals, plants, humans, spirits, objects, all of them of similar size, exceeding any particular relational context or frame and yet also always already intertwined with one another. The intimate enclosure of the decorated space of the mausoleum effects both a deepening and an opening of vision, then. The dim ambulatory circles a core of spacious brightness, and the visitor eventually reemerges into a world of both shadowy mystery and shining, bursting things.

The mosaics that adorn late ancient churches do not only wheel overhead but also unfurl underfoot. In many cases, floor mosaics are all that now remain to be seen of the original adornment, once roofs and walls give way and stones and tiles are dispersed—whether destroyed, buried, or finding their way into new structures. Here too, close to the earth and now exposed to landscape and sky, we discover distinctly "earthbound images." As Rina Talgam notes with respect to the rich remains in Israel and Jordan, "The most widespread decorative motifs in these floor mosaic are animals and depictions of rustic life and hunting, generally located within geometric or floral networks or, alternatively, scattered more freely over the surface of the floor."[101] Like the ambulatory vaults of Santa Costanza, these adorned floors raise questions that are difficult to answer definitively. Do their nature-oriented subjects convey a distinctly Christian message or are they merely decorative (whatever that might mean)? If they convey a distinctly Christian message, are they meant to be read literally or allegorically? Their placement further complicates matters. Whether meaningful or empty of theological significance, were these subjects chosen because they were considered suitable for treading upon? Documented opposition to the practice of inscribing crosses on church floors testifies that Christian opinion was divided regarding the appropriateness of walking on holy images.[102]

Let us assume, for the sake of argument, that images of divinely made creatures (as Christians would have perceived them) are likely to be encountered neither as meaningless nor as purely allegorical, least of all as intended for insult. How, then, would such images perform and relate within a basilical assemblage?

Consider, for example, the striking mid-sixth-century mosaics that cover the floors of the side aisles of a church at Petra, in modern Jordan.[103] As ever, our view of the larger ensemble of which they are a part is painfully partial. This constriction does allow, however, for a certain focus. After describing an apparently exquisite mosaic floor in the church of the Virgin of Pharos that depicts a variety of animals, the ninth-century Constantinopolitan patriarch Photios complains that the decoration of the church overwhelms the spectator, who "is carried and pulled away from one thing by another" (Homily 10.5). Contemporary viewers of ancient remains suffer from fewer such distractions. While reminding ourselves of how much more would have clamored for attention in a late ancient church, we may still profitably narrow our own attention to how a floor would have welcomed and affected a visitor.

The church is perched on a hill overlooking the main street of Petra; the hill itself lies at the end of a valley that is surrounded on both sides by craggy, cave-studded mountains that glow orange when the sun strikes them. Entering from the courtyard through the portal of the southern aisle, one is invited to step through the seasons, in the company of other creatures. Looking down may be easier on the neck than looking up, but one must pay attention to where one plants one's feet. "It is always as if one were treading on the other's toes," as Derrida puts it, speaking in more general terms.[104] Or simply treading on the other: as the decorated floor reminds us, we impinge on things

Figure 11. "Summer." Medallion floor mosaic, sixth century, Petra Church, Jordan. Photographer, Manuel Cohen/Art Resource, New York.

with every step we take. In so doing, we enter into relation with them, whether we are aware of it or not. And while it is possible to ignore the figures on the floor, the mosaic makes a strong bid for the viewer's attention. Walking down the middle of its three columns toward the front of the church, she encounters a series of anthropomorphic figures, facing in her direction. Winter, spring, summer, and autumn are depicted in female form and duly labeled (though the figure of winter has been severely damaged). Each frontally facing bust meets the viewer's eyes directly with an unwavering, round-eyed gaze. Summer is particularly arresting: voluptuous, she bares her right shoulder and breast, the latter positioned prominently in the center of the field, raising a scythe in her right hand while she holds a large fish against her body with her left arm, as if it were a human child (Figure 11). Other anthropomorphic figures are interspersed: earth and ocean, represented as female and

Figure 12. "Wisdom." Medallion floor mosaic, sixth century, Petra Church, Jordan. Photograph courtesy of the American Center of Oriental Research in Amman.

male, respectively, follow quickly on the heels of winter; they are followed by spring in turn, while *sophia*, or wisdom, also labeled as such, appears in the center of the long mosaic, followed by summer and autumn. In contrast to the unambiguously feminine figures of the seasons, *sophia* (whose face is unfortunately missing) is partly masculinized: her rounded breast modestly draped, the haloed figure holds a closed book in her left hand, which is covered with the edge of her robe; she makes the gesture of an orator with her right hand (Figure 12). Is this posture intended to be recognizably Christlike, evoking the divine son in his guise as incarnate word and wisdom?[105] Does the figure (also) invoke "secular" philosophy, or love of wisdom?[106] However wide the possible range of associations, here *sophia* clearly takes her place within the seasonal cyclings of a teeming cosmos. The final panels in the central column depict an eagle and two birds drinking from a goblet or basin of water; anomalously nonanthropomorphic, these figures apparently gesture toward the salvation promised to Christ's followers: the eagle may here as elsewhere signify resurrection;[107] the container of water, baptism, the baptismal eucharist, and/or Christ as the source of living water (cf. John 4:10). Right and left columns depict birds, fish, and other animals—goats, oxen, donkeys, lions—arranged symmetrically. Thus the aisle floor attunes viewers to the more-than-human world: even the anthropomorphic figures represent elemental forces, surrounding holy *sophia* as if she were a kind of cosmic queen, bearing the very book of creation.[108] The sole exceptions to this rule are the images of two fisherman and one fowler in the center column, humans shown in relationship to the creatures of air and water that dominate the right and left columns.

The north aisle mosaic is similarly arranged in three columns. In this case, instead of alternating medallions, rectangular panels, and conches, as in the south aisle, a series of identical medallions unfurls in a vine scroll pattern. Here a viewer walking along the central column would tread mostly on objects neither human nor animal; vases and basins holding liquids and baskets of fruits and vegetables dominate. These objects are flanked by a wide range of animals in the outer columns, wild and domestic, exotic and native, for the most part symmetrically arranged in facing pairs. The floor pulses with energy. Some of the animals break out of their containing medallions, as if on the verge of breaking free from the mosaic itself. Many of them are depicted in lively poses, caught in the midst of movement. A dog snaps at a bunch of grapes, a hare leaps, an ostrich prances, a bear roars, a rooster crows, a jaguar growls (Figure 13). In one case, figures interact between medallions, as one man ties a piece of wood onto the back of a camel while another holds it by its bridle. (These are among the few humans represented on this mosaic.) But for the most part, here as in the south aisle, as well as in some of the vault mosaics of Santa Costanza, animals and other things appear in isolation, floating in their bubbles without either spatial or temporal context, without even relative size: a bird is depicted on the same scale as an elephant.

These animals and objects present themselves as themselves, then, not as parts subordinated to a larger, hierarchically ordered whole. Yet at the same time, they arrive in and as fragments. Each is literally composed of fragments—small stone tiles patiently pieced together—and each creature itself appears as a fragment precisely in its combined isolation and proximity to the others. We recall Leslie Hill's suggestion that the fragment "intervenes as a perpetual opening to alterity."[109] As a field of such fragments, the mosaic constitutes a perpetually open zone of relationality. Men and camel are explicitly placed in relation to one another, but the others invite relation. According to symmetrical arrangement, animals of the same species face each other across

Figure 13. "Jaguar." Medallion floor mosaic, sixth century, Petra Church, Jordan. Photographer, Manuel Cohen/Art Resource, New York.

the central column: like may call to like, then. Juxtaposed difference is equally suggestive. What is the relationship of a carnivorous boar with gleaming tusk to the basket of fruit that it faces most directly and immediately? (Perhaps it is entertaining the possibility of vegetarianism.) Or to the two partridges below it? (Perhaps fowl would make a tastier dinner after all.) What is the relationship of the two partridges themselves to the sand grouse in the birdcage that separates them from their corresponding pair? How does their freedom speak to its confinement, their coupling to its solitude? And so on. The mosaic fragments also invite relation between the mosaic and its viewers, most of whom would not, after all, proceed or process down the length of the aisle purposefully but would, rather, join others standing or milling about in a worship service. Perhaps regulars would drift to a customary place near or upon a favorite, or at least familiar, figure; they would have a relationship to that figure, and they might wonder in turn (as I have) how that figure related to various ones around it.

Finally, the creatures in the mosaic invite relationship with the outside world, reminding visitors that there too we are always treading on others. The tesserae of which they are composed are likely cut mostly from local stone, famous for its vivid hues of brown and red;[110] thus, the adorned floor of the basilica is materially continuous with the living rock from which it comes and the sand to which it gradually returns. At the same time, the creatures that gaze up at us from under our feet point toward an expanse far wider in its imaginative scope than even the starkly beautiful desertscape of the church's immediate surroundings—a space mysteriously other, encompassing oceans and elephants, peacocks and giraffes, and fruits in abundance.

Wandering from Palestine to Italy and back again, we have encountered a column, a text, a vault, a floor. Collectively they help us apprehend late ancient church buildings as "interfolding network[s] of humanity and nonhumanity."[111] By participating in such relatively stable assemblages, Christian holy things—relics and images in particular—intensify their agency and affectivity. Sacrificing some of their mobility, they acquire new powers of attraction, coming into relationship with more things and people—touching and touched, feeling and felt not just by one but by a crowd. Does this increased promiscuity undercut the intimacy of the one-on-one relationship that an individual Christian might have with a fragment of wood or an image of a saint privately enjoyed? It changes the relationship, but it does not lessen its power. A column polished by the caress of a multitude of hands will only hum more vibrantly, and a mosaic burnished by the caress of a multitude of eyes will only gleam more brightly; crosses that call to each other will only speak more loudly of the one who died so that others might live. As animate collectivities, the buildings themselves "exert a force on the world," moreover.[112] They take others into their embrace and give them back again. They think out loud in images and speak silently in letters. They guide footsteps, direct eyes. They channel light and harbor shadows. Looking beyond their own doors, they communicate with one another, and with the landscapes that shape and are shaped by them. Enfolded and enfolding, they invite a sense of connection and communion. They also reach out to the widest world of which they can conceive—earth and ocean and starry sky; a cosmos that hums and gleams and calls; a mesh of lively, relational things, rising and falling with the temporal tide of the seasons. To adapt Bachelard's words, "Houses that are as dynamic as these allow us to inhabit the universe. Or, to put it differently, the universe comes to inhabit these houses."[113]

Interlude: Fragments of a Material Theology of Things

Things persist, by leaving traces of themselves.

Material remains: traces, relics, fragments. Ruins.

A lost ring, a toppled column, a small flask buried under shifting sand. Here a building intact, sheltering its treasures; there a few bits of stone wall, a floor, open sky. These things arrest my attention. They grab me.

The calm, still eyes of a woman meet my gaze across the chasm of their strange familiarity; so too do the eyes of bird, cheetah, fish, and more. Gold flashes: I look up. Now, the curve of a vase, the twist of a vine draws me. Image or thing? An image is a thing too—a thing, and a trace of a thing. "Suddenly," the thing (it *might* be anything) "reaches me. It animates me, and I animate it . . . : this is what creates every adventure."[114]

We have been searching for signs of life, for a way of living with attunement to the liveliness of things. "What is at stake is not beauty but an intensity that can be restored to us."[115]

But why search among ruins?

Incomplete, imperfect, ruins are no longer, or not yet, whole. They are in process—a process of disintegration, perhaps, or of reassembly, or of both at once. The process may move swiftly or so slowly it can scarcely be discerned. The scale may be simultaneously massive and minute.

Ruins lure imagination, enticing questions; they draw us to extend their incomplete thoughts: might once have been . . . ; is now . . . ; could yet be. . . . We become part of their incompleteness, they of ours.

All things arrive with a history—a history of relations and affect and sensation—and most of this remains unknown to us. Often we perceive things only in the melancholy moment of their disappearing, when they—and we too—have already begun to fall apart. Often things present themselves to one another in the very gesture of their withdrawal.

Or else: in the giddy moment of their arrival, before they—or we—have quite collected our selves. Still just bits and pieces!

Life emerges from amid the ruins after the end of the world as we know it. "Ruin seems something of the aesthetic texture of our future."[116] Perhaps it has ever been so. And: now, more than ever, it *is* so. The fear that haunts: "We've ruined the planet."

Things persist, by leaving traces of themselves. These are the relics. They are what remains, what survives, afterward.

"Today is not the end of history. We're living at the beginning of history."[117]

A relic is not an end but an opening onto a beginning after the end. It punctures closure. It invites relation. It proliferates promiscuously.

A relic is a thing that persists. It *is* the trace of the persistence of things, in their most radiant and adaptive vitality. It is life emerging amid the ruins.

Speaking Things: Rhetoric and Performativity in Basil's *Hexameron*

Things elude the very language that chases them. Yet words are things too—active, relational, and themselves elusive. Words are things that *do* things, as J. L. Austin so famously argued: they are things that *perform*.[118] One way that they perform is by conjuring other things, much as visual images can also do. These conjured things stand in mysteriously close relation to the words that bring them forth. Indeed, divine speech very nearly eliminates the gap between the two: "And God said, 'Let there be light,' and there was light" (Gen. 1:3). As the fourth-century Cappadocian bishop Basil of Caesarea remarks, "So, with an utterance, the maker [ποιητής] of all things all at once gave the gift of light to the cosmos" (Hexameron 2.7).[119] Here there is no sharp distinction between saying and shining; to speak a thing is to invoke its shimmering presence, "all at once." Fittingly, *poiēsis*, or making, designates the production either of a poem or of creation writ large. God as *poiētēs* speaks the poetry of creation, then. But it is Moses, with whom God uniquely communicates "mouth to mouth" (Num. 12:8), who embodies this speaking in and as scripture (Hexameron 1.1; cf. 2.7). Divine speech and biblical writing also stand in mysteriously close relation.

And what of those who revoice scripture, as Basil does? We may recall the pilgrim who enables a column to speak by reading its inscription aloud. So too Basil enables the Bible to speak by reading from its pages. In both cases, a visible word is translated into an audible one, as the enduring materiality of a written text—itself a kind of relic—takes on the more vivid but transient presence of performed speech. "In the beginning god made heaven and earth" (Gen. 1:1): these words constitute at once the first verse of the Bible and the first line of Basil's first homily on the *Hexameron*, or *Six Days of Creation*.

Thus, the Holy Spirit dictates, Moses writes, and Basil cites. Others cite too. Divine speech carries iterative force, then. Yet each utterance is also a singular event.[120] Basil's voice is distinctive, his rhetorical performance compelling. Preaching a series of nine sermons in the basilica of Cappadocian Caesarea in 378, he effectively reanimates the biblical text of Genesis 1:1–25.[121] In so doing, he opens his audience's eyes to the wonders of a world of vibrant things. The power of his words remains in play beyond this initial event, moreover, as those words are themselves transformed into a written text that may be rendered once again lively with each new reading. His friend Gregory of Nazianzus describes the effects of reading Basil's published sermons as follows: "Whenever I take his *Hexameron* in hand and quote its words [καὶ διὰ γλώσσης φέρω; literally, carry it with my tongue], I come to be in communion with the creator, and I comprehend the grounds of creation [κτίσεως λόγους], and I marvel [θαυμάζω] at creation more than before, when I used sight alone as my teacher" (Oration 43.67). Basil's sermons would once have resounded within the space of a church building, potentially yoking hearing with sight as they resonated with the images and other objects therein assembled. Through writing, they also have the power to travel beyond the church and its things. In so doing, they may bring traces of those previous material relations into new spaces, including spaces of imagination.[122] Basil envisions his sermons themselves as a temple entered through a forecourt and portico, harboring the "innermost shrine" of the holy of holies within (Hexameron 2.1). As we shall see, that verbal temple is richly adorned with images that draw the reader not only in but also out again, to "marvel at creation more than before."

Basil evokes wonder first of all by giving it rhetorical force in his own discourse. Having cited the first verse of Genesis for the second (but by no means the last) time, he comes to a stunned stop, without actually ceasing to talk. "Wonder [θαῦμα] at this thought halts my speech," he declares, marveling at the very fact of divine creation. His audience is invited to share his astonishment. "What shall I say first? Where shall I begin [ἄρξομαι] my narrative?"

(Hexameron 1.2). His own performance of the difficulty of beginning calls attention to the fact that the cited verse is itself both a beginning and an evocation of a beginning. "*En arkhē*," writes Moses: "*In beginning*, God made . . ." "It is an appropriate beginning," observes Basil, "for one about to narrate the formation of the cosmos to put a beginning to the ordering of visible things at the head of this discourse" (1.1).[123] He adds subsequently, "Perhaps these words signify the rapid and instantaneous moment of creation . . . , since the beginning is indivisible and without extension" (1.6). Basil's speech seems to hover over this elusive moment of emergence, as his discourse struggles to find its own beginning. He pauses at the threshold of creation, bemused and bedazzled.

Soon, however, time will have launched itself, in and through the text of Genesis as well as in Basil's own exegetical *logos*. Tracking the temporal unfolding of the initial, pretemporal *arkhē*, Basil can now tell his story, evoking yet more astonishment in his readers. "For even if we fail to understand the nature of transient things, nonetheless what is universally coming under our sensory observation evokes such great wonder [θαῦμα] that even the most capable mind appears less than the least of those things in the cosmos, with respect to the capacity either to fully examine their cause or to render due praise to their creator!" (1.11). His discourse remains highly performative, less describing past events than drawing readers into the temporal present of the biblical text. Having preached his first sermon in the morning, he delivers his second later in the day, ending with verse 5: "And the god called the light day and the darkness he called night. And evening [ἑσπέρα] came to be and morning [πρῶι] came to be, one day" (Gen. 1:5). Basil calls attention to the temporal convergence: "But the words about this [first] evening are overtaken by the evening that has arrived, thus putting an end to my discourse" (Homily 2.8). As he commences his third sermon on the morning of the second day, he notes similarly, "We are met together again on the second day to behold the wonders of the deeds of the second day" (3.1). Thus the time of creation and the time of Basil's marveling speech coincide.

Or rather, they *almost* do. In fact, his sermons on the six days of creation will take place over not six but five days. While his morning sermons are all delivered on the correct day, so to speak, three of his evening sermons—those delivered on days two, four, and five—anticipate the next biblical day's verses and themes. The anticipatory leap of day five in particular allows him to incorporate the sixth day of creation—dealing with earth-dwelling creatures—into his fifth day of hexameral preaching. Intriguingly, he seems to draw attention to the leap by initially exaggerating it. Having "forgotten" to tell of the air creatures, he begins to narrate the creation of the terrestrial animals prematurely, on the morning of the final day; as he corrects his omission by turning back to the birds, his narration of the terrestrial animals is subsequently deferred to evening—though it still comes too soon, with respect to biblical time. By thus packing (almost) six days of creation into five days, he clears space for a sixth day of preaching that is deferred—that of the creation of humanity, which is only anticipated in his final sermon. This subtle temporal rearrangement gives humans their own day rather than lumping them with the other terrestrial animals, as the biblical text does; yet that day does not quite arrive. The published sermons include only the five days, and those five days encompass all creatures but the human. The anthropocentrism of both Genesis 1 and its ancient Christian interpreters is thus bracketed, so to speak. Basil gives us a creation without humans, or at least without humans yet.[124]

Basil's sermons are not only performative in their lively citationality, their infectious affect, and their biblically synched temporality. They are also performative as reenactments of divine command. Basil calls attention to the fact that scripture shifts quickly from third-person narration—"In the beginning the god created" (Gen. 1:1)—to the direct speech of command—"Let there be light" (1:3). Moses did not have to make god "command and speak," but he did (Hexameron 3.2). As we have noted, divine command is performative speech par excellence, by definition wholly effective: the creation of the cosmos is no sooner said than done. Yet on Basil's reading, this performative agency, although apparently absolute and exclusive, turns out to be both dispersed and inclusive. Far from simply rendering other entities passive, the god's command—the divine word, or *beginning*—imbues them with agential force. To be commanded is to be urged to obey, and to obey requires agency: "Obedience cannot be the abandonment of freedom," as Karmen MacKendrick argues.[125] That agency is amplified, moreover, when the command obeyed is (perhaps inevitably) the command to *act*, in this case either simply by coming to be (γενήθητω) or by causing other things to come to be. In other words, the agency dispersed is the agency of genesis or creativity itself.

Who is commanded? Light, sky, waters, earth, sun, moon, stars. Of these, earth and waters play particularly important roles in the generation of the cosmos. We may recall the personified representations of earth and ocean on the mosaic floor of the south aisle of Petra's basilica, which seemingly echoes a biblical sensibility that also resounds through Basil's sermons. He cites the divine imperative: "*Blastēsatō hē gē*: Let the earth bring forth." Marveling that this "small utterance—this "brief command"—has such a powerful effect, so that what was "cold and sterile" now "labors all at once and is stirred to productivity," he compares the fertile earth to "one who has cast aside a sad and mournful garment and put on a brighter one, exulting in this proper adornment and holding the infinite variety of growing things before her" (Hexameron 5.2). Just so, late ancient mosaics like Petra's depict earth as a woman holding the edges of a cloak filled with fruit.[126] Earth's fecundity extends, moreover, to the fecundity of the plants that the god commands it to produce. "Let the earth bring forth the herb of grass generating seed according to its kind and according to its likeness, and the fruit tree producing fruit whose seed is in it, according to its kind on the earth" (Gen. 1:11): Basil leans into this biblical language, offering detailed descriptions of the wondrous generativity of grasses, shrubs, vines, and trees, which reproduce through both seeds and rhizomes, as he explains at some length. "'Let the earth bring forth,' he repeats, adding, "This small command was straightaway a vast nature, an artful system [λόγος]." Indeed, the command is a gift that keeps on giving, as the earth continues to manifest "its power to produce the generation of herbs, seeds, and trees" (Hexameron 5.10).

"What shall I say? What shall I leave unsaid?" Basil wonders (5.4). He does his best to say it *all*, as if the earth's very greening depended on his own recitation of botanical lists and lore. *Reed, couch-grass, mint, crocus, garlic, flowering rush* (5.2). *Wheat, oats* (5.3). *Hemlock, hellebore, monkshood, mandrake, poppy* (5.4). *Fir, cedar, cypress, pine, rose, myrtle, laurel, black poplar, willow, elm, white poplar, grape vine, olive tree* (5.6). *Pomegranate, almond, fig, palm* (5.7). *Apple, date, dogwood, turpentine, walnut* (5.8). All have their uses and their beauty. Plants provide food and medicines for humans and other animals; they also provide instructive exemplars. Grasses remind of the transience of all flesh (5.2); grapevines model the fruitfulness of souls and show how "claspings of love, like the tendrils of the vine, should attach us to our neighbors and make us rest on them" (5.6); and wild fig trees, lending their greater strength to the productivity of cultivated ones, teach Christians to discover sources of moral inspiration even among pagans and heretics (5.7). Surveying earth's sensuous bounty, Basil marvels at the extraordinary variety of shapes, colors, flavors, scents, and textures. Yet his words inevitably fall far short in depicting the diversity and teeming fertility of plant life. "What discourse can touch all?" he queries (5.8). Nightfall brings his overextended speech to a halt (5.9), but still so much remains to be said. "Our words never really speak except through an exhaustion of voice," writes Chrétien, "by being turned without cease to what exhausts the voice while promising it."[127]

Just as earth is commanded to generate living things, so too are the waters: "*Exagagetō ta hudata*: Let the waters produce," Basil cites. Specifically, the waters are commanded to produce "the moving creatures of the animals [ἑρπετὰ ψυχῶν ζωσῶν]" and "winged creatures flying in the firmament of heaven" (Hexameron 7.1; Gen. 1:20). Basil tackles the former group first: "The sea labors to give birth to all kinds of swimming things" (Hexameron 7.1). Once again he marvels at how much is accomplished performatively with so few words, while he also protests the impossibility of naming the vast number of oceanic creatures—even with an excess of words. "To review them all would be to undertake to count the waves of the ocean or to measure its waters in the hollow of the hand" (7.1). Still, he tries. *Seal, dolphin, ray, fish with skin and fish with scales* (7.1). *Muscle, scallop, sea snail, conch, oyster. Crab, lobster, polyps, cuttle fish, weever, lamprey, eel, sword-fish, cod, tuna* (7.2). *Squid* (7.3), *sea urchin* (7.5). *Coral, sea pinna, whale, shrimp, sprat, swordfish, sawfish, dogfish, shark, sea-hare* (7.6). Sea creatures may provide negative examples—such as "the cunning and trickery of the squid"—or positive ones—such as the generic capacity of fish to remain in the regions appropriate to them (7.3). They may not have the capacity to reason, but nature guides them truly in their migrations (7.4). "I myself have seen these marvels," Basil exclaims (7.5). Having, in effect, undertaken to count the waves, he declares that he has submerged his own discourse by invoking such wonders of creation, "coming one after the other in constant succession like waves" (7.6).

A night intervenes before Basil turns to the category of birds. "Why do the waters give birth also to birds?" he asks the next morning. The biblical text suggests affinities of movement: fish swim in the water, and birds float in the air (8.2). The variety of flying things is, once again, cause for wonder. *Eagle, bat, wasp, beetle. Dove, crane, starling, jackdaw. Rooster, peacock, partridge* (8.3). *Bee* (8.4). *Crane, stork, halcyon* (8.5). *Turtle-dove, osprey* (8.6). *Owl, night raven, nightingale, goose, vulture, swan* (8.7). "Imitate the distinctiveness of the bee," urges Basil, "which constructs its cells without injuring anyone and without interfering with the goods of others" (8.4). Storks are to be admired for their intelligence and their care for their old (8.5), bats for their love for one another (8.7). But these examples merely scratch the surface of nature's variety. "Think of all those whom my speech has left out, to avoid tediousness and not to exceed my limits," urges Basil. "Recognize everywhere the *sophia* of the god; never cease to wonder." Again, we may recall an image from the mosaic floor in the basilica at Petra, divine *sophia* represented as a woman holding a book, placed in the center of the "spectacle of creation," as Basil names it (9.5); the central column on this south aisle mosaic is flanked by eight pairs of sea animals and five pairs of birds (as well as six pairs of land animals), and there are three images that include birds in the central column.[128] Where the water creatures overwhelmed Basil like waves, the flying ones now outstrip him with their wings: "I remain further behind than I should if my feet had tried to match the rapidity of their flight" (8.7). It is as if he has only just begun to describe the creatures of the air, yet his sermon has run overtime; he has been carried away by his topic. However insufficient his account, he must make an end of it: "earth is calling me to describe" the land animals, he declares (8.8). Thus, even as Basil is reciting biblical words and thereby revoicing divine commands, he is also being commanded by the very earth of whom he speaks. Chrétien observes, "Every voice, hearing without cease, bears many voices within itself."[129] In Basil's listening response is the call of the god, of Moses, and of earth itself.

"*Exagagetō hē gē*: Let the earth produce." Like the waters, earth is doubly obedient and doubly fecund: having labored to give birth to plants, it "does not cease to obey the creator" (9.2), bringing forth "animals [ψυχὴν ζῶσαν] according to kind, quadrupeds, reptiles [ἑρπετὰ], and wild beasts of the earth according to kind" (Gen. 1:24). Along with plants and sea and air creatures, terrestrial animals have a share in the generativity of the earth and ocean that give them life: "The nature of things, put in motion by a single command, traverses creation through birth and death equally, and preserves the succession of kinds through resemblance to the last." Yet earth also continues to produce directly from her own flesh, so to speak. When grasshoppers, insects, mice, and eels emerge from dirt or mud (as Basil asserts they do), we hear the divine command reverberating still: "Let the earth produce an animal" (Hexameron 9.2). Once again, Basil calls the creatures by names—domesticated and wild beasts alike. *Ox, ass, horse, wolf, fox, stag, ant, dog, lion, panther, bear, tortoise, sheep, hedgehog, wolf* (9.3). "All bear the marks of the *sophia* of the creator" (9.4). *Hare, goat, viper, camel, tiger, elephant, mouse, scorpion.* "It is not only in large animals that we see unsearchable *sophia*; but even in the smallest no less wonder is evoked" (9.5). Each animal has its own distinct characteristics and each has a lesson to convey. As Patricia Cox Miller suggests, "Basil's handing of the earth's creatures" is often oriented quite explicitly to the molding of human subjects, modeling "a therapeutic use of animal-pictures to craft an ascetic self."[130] As Basil puts it, "Beasts are witnesses of faith" (9.6).

In closing his last sermon, Basil protests his reluctance to take up the topic of the creation of humanity. He imagines his audience clamoring to hear just that. They complain, "We are being taught about the nature of what is ours [τὰ ἡμέτερα], but we are ignorant of ourselves" (9.6). *Ta hēmetera*, or "our things," may indicate possessions or property, but it may also suggest a sense of belonging or connectivity. These *things*—sky, water, earth, plants, fish, birds, animals, and more—are not us but they *are* ours; conversely, we are theirs. It is difficult, even impossible, to know the other creatures: both time and words fall short. But it is even more difficult to know ourselves, asserts Basil, defending his own hesitancy to proceed while extending his strategy of deferral with an anti-Arian theological detour. Once again, he is saved by the very circling of days that his discourse performs: "But evening, which long ago sent the sun to the west, imposes silence on me," he declares. "Here, then let me be content with what I have said, and put my discourse to bed" (9.6). Putting his discourse to bed before giving an account of the creation of humanity, he offers a knowledge of the human not in itself but in relation to all those others to whom it belongs.

Or rather, in potential relation. Like the Petra mosaics, Basil's nine sermons on the six days of creation present creatures in all of their variety and distinctness, subordinated to no particular order or hierarchy of relations; differences of size or scale are ignored, explicitly so. Orders and even hierarchies may be insinuated—nonpoisonous versus poisonous plants, for example—but they are swiftly deconstructed or made to give way to different orders.[131] What is poison to one animal may be food or medicine to another, or even to the same animal if ingested differently or in a different quantity. And plants may also be sorted differently—as seed-bearing versus rhizomatic, for example. We are presented, then, with a "democracy of objects," astonishing (Basil insists!) in their number, variation, and individuated perfection, and open to limitless possible relations. The *Hexameron* performs the cosmos as an abounding, generative excess of *things*, an ongoing work of divine *poiesis* that remains ever emergent, unfinished, fragmentary.

It is fragmentary most explicitly and pointedly in its deferral of the introduction of the human, as we have seen. Basil's final sermon hovers on the threshold of Genesis 1:26. The first half of the verse is read, if not fully interpreted: "And the god said, 'Let us make the human according to our image and likeness." The second half, however, is not yet read: "and let them have dominion over the fish of the sea, and over the flying creatures of the sky, and over the cattle and all the earth, and over all the reptiles that creep on the earth." A similar deferral can perhaps be observed in the floor mosaics of the Petra basilica. There vine scrolls twine around figures of fish, birds, and land animals, as well as vases and basins and baskets containing water and the fruits of the earth. Humans appear only as personifications of earth, ocean, the seasons, and cosmic wisdom or, in rare cases, in interactions with other animals or things. Humans as such appear only in relation, then. Admittedly, those relations are already tilted toward dominion: fishermen and fowler track their prey; camel handlers subject a beast to its burden; and slaves carry vases in a domesticated scene of human-to-human subjugation. Yet there is a tentativeness as well in the dominion thus depicted: to put it simply, humans take up very little space, nor do they seem, finally, to succeed in fully controlling the terms of their own relations. This is especially evident in the depiction of the camel and its human handlers, which sprawls across three medallions, as we have seen.[132] In the center, the camel is in the process of either lying down or standing up, its head protruding into the right medallion—an unstable and suspenseful moment. On the right, one man faces the camel, holding its bridle and bending toward it; he is not exactly leading it, then. On the left, another man strides behind or toward the camel, holding ropes looped around the large piece of wood that is strapped to its back. Is he in the process of securing the wood? Unloading it? Using it to tether the camel? And what is the man in front doing? Ambiguity creates tension and uncertainty, as if the camel's indeterminate movement has thrown all of the interactions off balance. When humans enter the picture, they do not simply call all the shots, then.

An ivory plaque, perhaps carved at Rome circa 400, may capture the moment and mood of Basil's *Hexameron* even more effectively than the Petra mosaics do. Tradition gives it a leading title, "Adam Naming the Animals," but we should perhaps hesitate before following this lead. The first human floats ambiguously near the top of a vertical panel whose space is otherwise filled with a rollicking panoply of animals. Adam's gaze is unfocused, his mouth closed, and he lacks visible ears with which to hear the roars, growls, and grunts of the contrastingly open-mouthed beasts. His left hand holds on to the branch of a tree and his right hand reaches downward. The first two fingers of the latter point in what might be construed as a typical orator's gesture, but the hand is not raised and may just be supporting the human on the rock on which he is perched somewhat unsteadily.[133] In short, it is not at all clear that Adam is doing any naming on this ivory carving. Indeed, like a newborn infant, he may have only just arrived on the scene, as Glenn Peers proposes.

Suspended in the open moment of humanity's genesis (in the middle of the verse of Genesis 1:26, say), the plaque "will never allow Adam to master his environment, it will forever hold off the anthropozoic (. . . the incarnational era of divine approval of Christian-humanity's dominance) and forecast Anthropocene . . . to come," Peers suggests evocatively. "The outcome is forever ambivalent. . . . Animal precedes human here, and one waits in uncertainty to see what Adam makes of this moment when he takes possession (of himself)."[134] Just so, in Basil's *Hexameron*, animal precedes human, and we wait with uncertainty to see what humans will make of the others—*ta hēmetera*—and thus of themselves, in relation.

To return to the open moment invoked by Basil's performative text is not, finally, to deny the reality of subsequent histories of human dominion but rather to ask what might still be possible. It is to respond, ever and again, to the divine "call to be" that is ongoing, at once "eternal and instantaneous."[135] Like the Petra mosaics, and perhaps other images that may once have adorned the church where Basil preached, his sermons on the six days of creation evoke wonder at the startling liveliness and infinite variety of created things. Answering the call that resounds through materiality itself, these things also call to one another—to all their others. Never fully subordinated to a larger order, they are always inviting relation. Humans respond belatedly, or some of us do: the anthropocene is the haunting of that belatedness. But now is always the only moment we have. What might we make of it?

Desiring Things: Contemplation, Creation, and God in Augustine and Pseudo-Dionysius

Is it possible to love *all* things? Even to love all things equally? No one can. An ascetic practitioner as ambitious and optimistic as Evagrius of Pontus acknowledges that "it is not possible to love all the brothers to the same degree" (Praktikos 100); add sisters and nonhuman animals and things in all of their infinite variety, and the challenge is surely magnified. Yet just this impossible ambition drives Christian practices of prayer and contemplation nonetheless, according to some of its most brilliant late ancient theorists—not least Augustine of Hippo and the early sixth-century theologian known as Dionysius. For these thinkers, love of all things is crucially mediated by love of god. It may even be that the two loves are (very nearly) one and the same—that to love all things is to love god, and vice versa. Alternatively, it may be that god is not object but verb, that god *is* the loving of all things—more precisely, that god is *all things loving all things*. However it is parsed theologically (and this is a question to which we will return), to orient oneself toward such an impossible possibility as all-love is to undertake a radical transformation of attention, sensation, emotion, and thought. This is not the effort of individual humans, however: it is the collective work of ritual and the gift of participation in an expansive communion.

To be sure, the two loves, of all things and of god, also often seem to be at odds. In his famous theological autobiography *Confessions*, Augustine declares, "If bodies are pleasing, praise god for them, and turn love [*amorem*] back on their maker, lest you displease in what pleases you. If souls are pleasing, let them be loved in god" (Confessions 4.12). The problem here is transience. No thing is eternal. And no thing can bear the full weight of human desire, therefore. Love of transient beings is always haunted by grief, and humans regularly try to evade that grief; in so doing, they make false idols of what they love, denying their transience. Such idols are doomed both to disappoint their lovers and to displease god, suggests Augustine. Creaturely things arouse pleasure, and rightly so, but we should not cling to them as if they were stable and lasting. "All things rise and set, and rising they begin as if to exist, and they grow and they are perfected, and being perfected they grow old and perish. . . . They are parts of things, which do not all exist at the same time, but going and coming they all bring forth the universe of which they are a part" (4.10).

Sound illustrates this flux particularly well. "For even our speech [*sermo*] will not be whole if one word does not give way when it has sounded its parts, so that another may succeed it." Praise of god is not excepted from such temporal dispersion. Augustine recites, or perhaps sings, the first line of Ambrose's hymn, "*deus creator omnium*," in celebration of the beauties of creation. However, he begs his god that "the glue of love" not bind him too tightly to the vibration of bodily sensations, which dissipate so quickly (4.10). Later he will squirm with ambivalence about the power of music to move his soul, even bringing him to tears. He can only hope he does not enjoy it too much. "Now I confess that I find some rest in the sounds that your words"—that is, the Psalms—"animate, when they are sung with soft and artful voice." He adds hastily, "Not so that I am held fast but so that I may arise when I want." Anxious about the power of music to captivate, holding its hearer in thrall, he wonders whether it would be better to follow the advice of Athanasius, eliminating melody from the recitation of the Psalms altogether, so great is the danger of the pleasure it brings.[136] On balance, however, he thinks that the custom of singing should be retained, "so that through the delights of the ears a weaker soul might rise to the love of devotion" (10.33).

What is the solution to this predicament of desire, which is called to extend itself infinitely toward the divine, yet must also (it seems) observe limits with respect to the very sensations that arouse it? Augustine has told us: to love god through bodies, and to love souls—the liveliness of things, that is—in god. A subsequent reference to Ambrose's hymn helps unpack this enigmatic proposal, even if it takes us along an indirect route. *Deus creator omnium*, Augustine quotes again: "God, creator of all." He carefully analyzes the meter of this brief verse—iambic tetrameter, four sets of short syllables followed by ones twice as long. But how can he know that the long one is long in comparison, if the short syllable has passed away by the time the long one is sounded? For that matter, how can he measure the long one itself since it too has fled the present by the time he performs his measurement? The answer, he concludes, is memory. Memory holds together what is otherwise dispersed (11.27). Yet memory does not function alone. When Augustine recites a psalm, for example, his mind performs not one but three temporalizing acts: "It anticipates, it attends, and it remembers." Thus through expectation as well as memory, he holds the whole psalm together as he chants or sings, syllable by syllable, in the suspended moment of present attentiveness (11.28).

The time created by incantation is not strictly linear. Perhaps it is akin to what Jean-Luc Nancy calls "sonorous time." "Sonorous time," he writes, "is a present in waves on a swell, not in a point on a line; it is a time that opens up, that is hollowed out, that is enlarged or ramified, that envelops or separates, that becomes or is turned into a loop, that stretches out or contracts, and so on."[137] Such an undulating temporality[138] can partly overcome the dispersive "going and coming" of transient things. This overcoming is the work of liturgy—here specifically, of the responsorial singing of the Psalms,[139] divine words embodied in the melodious voices of humans. The swell of a song may hold even a universe of fugitive fragments together. Yet we only perceive the whole of it through the foresight of anticipation and the hindsight of memory, and then only fleetingly, as the last note fades. Augustine's is, after all, an always already fallen world, in which love inevitably comes too late. He fears the power of music even in the song that sings god's praises. Yet he himself seems also to suggest that it is precisely in experiences of such powerful sensory attunement that the boundaries between the sensory world and the divine, time and eternity, begin to dissolve.

I have emphasized the importance of sound for Augustine's theory of desire, anchored in the weight he places not only on psalmody but also on scripture more generally as a mediator of divine grace. Nonetheless, desire, in all its ambivalence, works across the sensorium, as Augustine depicts it. He is enticed by taste, touch, and smell as well as sound (10.6, 10.30–33); and he is enticed by "the pleasure of the eyes," as he tells us, parading his temptations before the reader. His god is his ornament (*decus*), but he acknowledges that other kinds of ornamentation or art also derive from that god: "Many beautiful things [*pulchra*] come from the beauty [*pulchritudo*] that is above, drawn through souls into artful hands" (10.34). The beauties of the natural world draw him most powerfully, and he performs his own engagement with that world as a kind of sensory call and response. As if taking a quick sampling from the created things named in Genesis 1, he seeks the divine in earth and sea, in the deeps and in creeping animals, in the air and the heavens, and in sun, moon, and stars. Are you god? he wonders, as he attends to each in turn. Each responds to him: I am not. Still Augustine persists in his inquisitive attentiveness. Finally, he comes to understand. As he tells it, it was through their very beauty that "they cried out with a loud voice, 'god made us!'" (10.6). In this "loud voice," at once one and many, we should not of course hear talking elements or animals in any literal sense but rather what Karmen MacKendrick describes as "a body that breathes song," borne on the divine spirit that enlivens it—"the breath of each material, animal, human body speaking its part in making the meaning of the world."[140] Here the physical senses as normally experienced defer to another kind of sensing or listening. Attention deepens. Then the world of things that draws Augustine in so powerfully lives and breathes and even has a kind of voice. But it also retains its self-exceeding mystery. Is god one name for that mystery?

Augustine's thought may be well described as panentheistic: all things are in god and god is in all things. He tracks the divine by tracking his own desire. "What do I love when I love you?" he asks his god, answering in his own voice: "Not the beauty of a body." But he adds, "And nonetheless I love a certain light and a certain voice and a certain fragrance and a certain taste and a certain embrace, when I love my god, who is light, voice, fragrance, taste, embrace" (10.6). It is through his senses—thus, presumably through created things—that god ultimately seduces him, with near violent forcefulness: "You called and shouted and broke through my deafness; you flashed and shone and banished my blindness; you shed your fragrance and I drew breath and now I pant for you; I tasted and now I hunger and thirst; you touched me, and I have blazed up in your peace" (10.27). Yet the senses also always threaten to draw him away from god.

Commentators have explained this apparent paradox by sharply distinguishing physical sensation from spiritual or intellectual sensation, a kind of "seeing" (*theoria, contemplatio*) that fully transcends the physical; and there is support for this in Augustine's writings, as in other neoplatonic thinkers. John Peter Kenney posits that "intellectual vision" remains for Augustine the highest form of contemplative perception and is, moreover, "clearly delineated from either sensory vision or the attenuated version of it that he calls spiritual vision."[141] However, Kenney also argues that Augustine ultimately disavows the soteriological efficacy of contemplative vision based on its disappointing transience, while affirming its cognitive or intellectual usefulness. Contrary to the assertions of the Platonists—Plotinus and Porphyry in particular—contemplation is not an end in itself but must, according to Augustine, give way to confessional self-reflection and eschatological hope, through a process in which the soul comes to grasp simultaneously its own abject weakness and its creator's potent grace. This inadequacy, or need for a supplement, characterizes Augustine's postbaptismal Ostia vision (Confessions 9.10), shared with his mother, as much as it does his previous "Platonic" ascent (7.10–7.20), on Kenney's reading. And here Augustine aligns himself (whether knowingly or not) with Porphyry's rival Iamblichus and other neoplatonists for whom "the true means to salvation was 'theurgy,' the practice of pagan ritual designed to effect the soul's return to a higher realm."[142] Confession itself becomes a kind of metaliturgical act, encompassing and transforming contemplation within a performance of prayer that not only appropriates the voice of scripture—especially the Psalms, as we have seen—but also offers a meditative interpretation of its first verses, thus also of creation itself. Put otherwise, Augustine turns from a Platonic practice of contemplation that is interior and intellectual (as he understands it) toward a Christian practice of contemplation that is at once exterior and interior, sensory and more than sensory—embodied and then some. If his contemplation of creation leads to a contemplation of scripture, scripture returns him to creation. Even Kenney may underestimate the distance that Augustine places between himself and a striving for strictly noetic contemplation—a distance measured not only by his particular understanding of divine agency and creaturely passivity but also by his strong affirmation of the abidingness of materiality.

By joining contemplation to confession and embedding confession in psalmody and prayer, Augustine securely "nests it within the church" and its sacraments.[143] Elsewhere he famously defines sacraments as the visible "marks of divine things [*signacula . . . rerum divinarum*]" in which "the invisible things themselves [*res ipsas invisibiles*] are honored" (On the Catechizing of the Uninstructed 26.50). This links his sacramental theology to his larger theory of signs: "A sign [*signum*] is a thing [*res*] that, beyond the appearance that is pressed upon the senses, from itself makes something else come to mind; just as when we see a footprint we think that the animal whose footprint it is has passed by" (On Christian Doctrine 2.1.1). *Things*, whether acts, objects, or animals, always exceed their sensory impressions; yet it is through those impressions or footprints that we can know them at all. Moreover, things refer us not only to their own surplus—their eschatological dimension, let us say—but also to other things with which they are in relation; their relations are part of their surplus. All things are signs, and all signs are things. A mosaic is no different from a mountain in this respect. For Augustine, sacraments, scripture, and creation itself are all networks of sacred, signifying things. Pointing to their own mysterious depths, things also point toward their divine author—to the signifying power that inheres in materiality, perhaps. Or, to invoke MacKendrick's words again, to "the breath of each material, animal, human body speaking its part in making the meaning of the world."[144]

That contemplation for Augustine is never detached from desire or sensation is confirmed by his attempts to imagine the end of things—creation in its resurrected plenitude, when "the beauty of the completed course of time . . . shall be finished, like the grand melody of some ineffably wise song master" (Letter 138.1.5). At the close of his *City of God* Augustine speculates on what life lived in resurrected bodies might be like. The apostle Paul promises, "Now we see in a mirror dimly, but then face to face" (1 Cor. 13:12). Augustine wonders if this face-to-face seeing will involve "bodily eyes"—the eyes of the spiritual body, that is. "An animal body is sown, a spiritual body rises," as Paul proclaims (1 Cor. 15:44). Augustine is inclined to think that the spiritual body, like the animal body, will indeed use its eyes to see incorporeal as well as corporeal things. He acknowledges "the reasoning of the philosophers," according to which "intelligible things are seen by the sight of the mind and sensible things (that is, bodily) by the sense of the body." Scripture suggests otherwise, however, he insists. Augustine speculates that the vision of spiritual eyes might be something like our current ability to sense not only visible bodies but also the life force (*vita*) invisibly animating them. Spiritual eyes will have a kind of double vision, then. "Thus wherever we move those eyes of our bodies, through bodies we shall gaze upon the incorporeal god ruling all things." He stresses again that this is not intellectual vision but an enhanced or transformed bodily vision. As he imagines it, "God will be known and visible to us in such a way as to be spiritually seen by each one of us in each one of us, seen by the one in the other, seen in him or her or itself, seen in the new heaven and the new earth, seen in the whole creation as it will be, seen also through bodies in every body, wherever the eyes of the spiritual body are directed with their penetrating gaze" (City of God 22.29).[145] The thoroughgoing incarnationalism of this theological assertion is striking: to see "face to face" is to perceive god *embodied in creation*—fully and all at once. Conversely, to perceive the beauty of bodies in their plentiful excess, in the fullness of their resurrected glory, is to see god. "There we shall rest and see, see and love, love and praise," concludes Augustine. For what does desire ever desire but to reach and rest in such a saturated state of mutual apprehension and adoration? Margaret Miles sums up Augustine's position as follows: "Spiritual vision relies on physical vision. . . . Moreover . . . the accurate 'seeing' of physical objects irreducibly involves the exercise of spiritual vision; to see accurately is to see lovingly, to participate in the very substance of the God-who-is-love."[146]

Admittedly, Augustine's thought is sometimes limited precisely where we want it to open up. In his description of the eschatological communion, he restricts his language of sensation to the visual; his own gaze appears to encompass only resurrected *human* bodies, and those bodies conform to a hierarchical order seen to be modeled by the human body itself (cf. 1 Cor. 12:12–31). He is sure that there will be "degrees of honor and glory" in the city of god, though this will cause no resentment or discontent (City of God 22.30). In all these ways, Augustine's thought remains conventionally Platonic, as well as Pauline. We recall, however, his strong attunement in his *Confessions* to the power of creation to seduce through all the senses and thereby (potentially) to draw us toward the divine. In that context, it becomes difficult to understand why Augustine should not finally attribute the pleasures of mutual touching, tasting, smelling, and hearing, as well as seeing, to resurrected bodies, or at least to human ones.

But what of nonhuman bodies? Augustine perceives beauty in all things. Is the feeling mutual? "Behavior and display go beyond sheer survival," notes

Morton. "Chimps paint and do rain dances. Perhaps nonhumans are capable of aesthetic contemplation, enjoying things for no reason."[147] Animal bodies contemplating human ones unsettle human self-certainty, moreover, not least because of the enigmatic character of their regard. Jacques Derrida explores at some length the puzzling shame he feels when caught naked in the gaze of a small cat; he tries to imagine a time before this shame, "when Adam . . . called out the animals' names *before* the fall, still naked but before being ashamed of his nudity."[148] Jean-Christophe Bailly too mediates on the force of the wordless gaze of animals. "The pensivity of animals," he writes, "establishes . . . that the world in which we live is gazed upon by other beings, that the visible is shared among creatures, and that a politics could be invented on this basis, if it is not too late."[149] A politics, a polity: a *civitas dei*, perhaps.

Vision may be shared by many animals, but the tactile sense is arguably shared by all of them; Bailly invokes "the close contact, always singular and always consisting of touch, that is the ordinary mode of the bond between them and us—something scarcely formed, always nascent."[150] Perhaps we can expand the boundaries beyond even the animal world. If touch is "the most fundamental and universal of all the senses," if it "unlocks for us and brings to us the root qualities that constitute every body as such," as Chrétien writes,[151] is its reciprocity limited to humans and other animals? My foot touches the ground, leaving an impression, and also feels its touch; if I am wearing a shoe, shoe and ground touch; if I tread on a dandelion, it may be crushed. Does the ground feel my foot, do shoe and ground feel each other, foot and flower too? Why not? "Things are connected to and flow into other things, always transforming and being transformed."[152] Affect is "not specific to humans, organisms, or even to bodies."[153]

Touch might be described as a mode not only of sensation but also of intellection that is shared by all things. That is, touch might be a place where the distinction between intellection and sensation begins to break down. To know is to touch something with our minds, to have our minds touched. When Augustine describes his experience of apprehending divine wisdom in his mother Monica's company, he says, "We touched it lightly for a full heart's beat" (Confessions 9.10). "You touched me, and I have blazed up in your peace," he tells his god (10.27). As this passage reminds us (should we need reminding), the pleasures of touch are among the most intense and most all-encompassing. So too is the pain to which touch exposes us. Paradoxically, the touch of another—animal, vegetable, mineral, or divine—confirms our sense of self as other while also seeming to dissolve boundaries through the ruse of immediacy. Touch may be violent, seductive, soothing, supportive, and more; but it is never a one-way street. Substituting touch for vision in Augustine's eschatological "city of god" might, then, yield a more expansive and dynamic communion of things. Yet the senses continue to spill into one another. Chrétien captures the insight of mystics: "When the entire body radiates and burns through this divine touch, it becomes song and word. Yet that which it sings with its entire being, collected whole and gathered up by the Other, is what it cannot say, what infinitely exceeds it—excess to which touch as such is destined, and which in the humblest sensation and least contact here below was already forever unsealed to us."[154] In Augustine's words, "Surely nothing will be more pleasant in this city than this song to the glory of the grace of Christ" (City of God 22.30). The glory of Christ is also the glory, the shining, of all things—the divine incarnate in creation.

Writing roughly a century later and at the other end of the Mediterranean, Dionysius shares with Augustine an intellectual vision strongly shaped by both Pauline and Platonic traditions. Like Augustine, he understands desire as the moving force in a cosmic liturgy through which creatures come to contemplate god in creation and creation in god. Significantly for our purposes, Dionysius's view of salvation is profoundly cosmological, encompassing the breadth of creation in its more-than-human scope. His treatment of hierarchy is, moreover, particularly well developed, allowing us to return to this vexed topic.

"One must hymn [ὑμνητέον] the beneficient providence of the thearchy from all created things [ἐκ πάντων τῶν αιτιατῶν]," states Dionysius in his treatise *On Divine Names* (5).[155] His language is, typically, both condensed and somewhat cryptic. *Pronoia* or providence names god as the overflow of ecstatic *eros* that creates the cosmos, in and as a process of divine incarnation and self-revelation.[156] *Pronoia*'s creative beneficence calls forth a song. The song is a cosmic song in two senses: it takes its start from the contemplation of the things created, and it arises from among created things. This singing is also a naming of the god, a naming that responds to and mimics the god's own creative outpouring. For the divine is hymned not only "from all creatures" but also "from every name" (On Divine Names 6); that is, the god is praised with the name of every creature, through the voices of all creatures. The Bible is the book of creation and the archive of divine names. Dionysius thus affirms the *via positiva*, or kataphatic approach: "and from all beings whatsoever she is harmoniously hymned and named" (7). The divine is not only called by the name of all things; she "*is* all things"—"sun, star, and fire, water, wind, and dew, cloud, archetypal stone, and rock." But if the kataphatic naming responds to and reenacts the god's creative overflow, as manifest in both cosmos and scripture, a corresponding apophatic unsaying responds to and reenacts the ultimate elusiveness of both god and creation. The divine is not only named by every name; she is also nameless. The divine not only "is all things" but is also "none of the things" (6). That is to say, the name of god is all names; yet even the totality of all names falls short. God both is and is not identical with the sum total of the created things that sing god's praises by calling her by all their own names.

Praise is carried on the waves of longing. "For all things both surround [providence] and are directed toward her, and she is before all things, and all things cohere in her. . . . All things long for [ἐφίεται] her." Different kinds of things—human, animal, plant, mineral—long for the divine differently, Dionysius goes on to explain. However, each desires in the way appropriate to it: "Intellectual and rational things, cognitively; things lower than these, sensually; and the rest, according to their vital movement or the fitness of their being and habit" (5). Complex beings like humans may long in all these ways: as Eric Perl notes, there is "a continuum of modes of communion with God" and "the higher types of participation do not exclude the lower."[157] To long, in whatever way, is to participate in divine love. As John Rist sums it up, for Dionysius, "God is Eros and the cause of Eros in all other things."[158] At once lover, beloved, and love itself, god is, in Dionysius's own words, "erotic motion . . . pouring out from the good onto all that is and returning once again to the good . . . , always proceeding, always remaining, always being restored" (4). Thus creaturely desire not only responds to but is continuous with divine eros, completing its circuit, so to speak; it is the overflow of cosmic eros turning back toward its own elusive source.

That flow is also a connectivity. Dionysius claims that his teacher Hierotheos wrote a work entitled *Erotic Hymns*, and he cites from that work as follows: "When we talk of eros . . . we should think of a unifying and comingling power that moves those higher to the care [πρόνοια] of those more needy, those of equal rank to mutual support, and finally those subordinate to a return to the stronger and the superior" (4). It is fitting that Dionysius's musings on eros emerge as glosses on fragments of an otherwise absent text, either lost or (more likely) fictive. As Sean Braune notes in another context, "The fragment itself is a remaining trace of a larger writing that exists as an imaginary supplement; as a potential writing once written and now lost, only existing as pure potential in a sort of libidinous energy catalyzing in the mind of the reader and critic."[159] Like the cosmic hymning of divine providence, the *Erotic Hymns* exist only as an imaginary supplement to Dionysius's own text, a sonorous plenitude as evocative and alluring and ultimately elusive as the god itself.

How inclusive is the "all" of creation that longs for god and sings god's praises by invoking the overflowing fullness of divine names? Addressing this crucial question requires considering Dionysius's understanding of both sacrament and hierarchy. Like Augustine's, Dionysius's cosmic hymnody stands in both close and ambiguous relation to his sacramental theology, a topic that is the focus of his treatise *On the Ecclesiastical Hierarchy*. The sacraments include baptism, eucharist, and ointment, as duly administered by the three hierarchical orders of the clergy; their aim is to ensure "the salvation and divinization of all reasonable and intelligent beings" (On the Ecclesiastical Hierarchy 1.4). Thus sacramentally mediated salvation seems to be available to humans alone.

The eucharist, which Dionysius refers to as communion or gathering, is "the sacrament of sacraments," a prime site of theurgy, or the work of god; "it draws our fragmented lives together"; "it forges a divine unity out of the divisions within us" (3.I). Dionysius is attentive to the choreography as well as the soundtrack of this sacrament. He describes how the hierarch, or bishop, begins the ritual with a prayer at the altar and then moves out from the altar, circling the sanctuary with a censer and returning thereafter to the altar; in this "double movement" he performs the divine dance of erotic procession and return, differentiation and unification, evoking a wider cosmic flow (3.III.3). Back at the altar, Dionysius explains, the bishop sings the psalms "and the entire assembly joins him in this"; the reading of scripture follows. As Dionysius comments, "the sacred psalmody is a part of the hierarchic mysteries and should certainly accompany the most hierarchic of them all," while the readings impart edifying lessons. In particular, the psalms offer "a universal song and narrative of all divine things and they enable everyone who participates in a godly spirit always to receive and to pass on the sacrament of the hierarchy" (3.III.4)—to participate in the "erotic movement" of the cosmos, that is. In Dionysius's view, the psalms do crucial work not only in preparing human spirits to receive the elements of bread and wine but also (and more immediately) in preparing them to receive the "images and proclamations" of other parts of scripture. With respect to the former, psalmody attunes its singers to the "divine harmony" and brings them into positive relation "not only with divine realities but with our individual selves and with others in such a way that we make up one homogenous choir of sacred humans." With respect to the latter, psalmody offers a "summary and opaque outline" on which the rest of scripture expands with more accessible images (3.III.5).

Following the readings, the deacons and priests lay out the sacred elements of bread and cup and the music continues, as the congregation now sings "the hymn of universal faith"—a possible allusion to the chanting of the creed (3.II). Dionysius adds, "This hymn is sometimes called a confession of praise, sometimes a symbol of adoration, sometimes . . . a hierarchic thanksgiving. . . . To me it seems that this song is a celebration of all the work of God on our behalf" (3.III.7). That divine work or theurgy—itself another echo of the erotic procession and return mobilized by human ritual—is concentrated in the sacred act "when [the hierarch] uncovers the veiled gifts, when he makes a multiplicity of what had originally been one, when the distributed sacrament and those receiving it are made perfectly one, when a perfect communion of all the participants is achieved" (3.III.13). At the close of the rite of communion, bishop and congregation "end the ceremony with a sacred thanksgiving" (3.III.15).

Hymnody thus suffuses and carries the eucharistic liturgy; sound is the medium of sacrament, though not of course the only one. It is perhaps not irrelevant that Dionysius seems to have lived in a time—the late fifth or early sixth century—and a place—Syria, perhaps—in which hymn writing was on the rise. An anonymous *Life of Severus of Antioch* (465–538) reports that "the Antiochenes rejoiced in song," so much so that Severus "appointed chanters and composed hymns"; his hymns were "full of sighs and calling those who hear them unto the tears loved by God" (Life of Severus 60).[160] In his treatise *On the Celestial Hierarchy*, Dionysius lets us hear an angelic choir echoing through the human song. "Theology has transmitted to the people of earth those hymns sung by the first ranks of the angels." He alludes to two hymns in particular, emphasizing their resounding loudness: "Blessed be the glory of the Lord of many waters" (Ezek. 3:12, LXX) and "Holy, holy, holy is the lord of hosts; the whole earth is full of his glory" (Isa. 6:3). The former as well as the latter is associated with the Sanctus, that is, the eucharistic prayer, from the early fourth century,[161] and such passages citing angelic song appear in other hymns of the period as well. As with the book of *Erotic Hymns* that he attributes to his teacher, Dionysius makes it clear that these two fragments merely gesture toward a larger, albeit imaginary, hymnic whole: "In my book *Divine Hymns* I have already explicated, to the best of my ability, the supreme praises sung by those holy intelligences which dwell beyond in heaven" (On the Celestial Hierarchy 7.4).

If humans receive divine hymns from the angels who stand in the closest possible proximity to god, do their own liturgically performed hymns and sacraments not overflow the merely human communion in turn, extending salvation to even the simplest creatures? "God is providentially available to all things and becomes all things in all for the salvation of them all" (On Divine Names 9.5). For Dionysius, ecclesiastical sacraments, while distinctly human, are embedded in a broader communion: all of creation signifies, pointing beyond itself; and all of creation yearns to be made divine. As Perl puts it, "The cosmos is sacramental."[162] All things inspire song, as we have seen, and all things sing. They sing songs of desire, erotic hymns of which most of us can hear only fragments. It is easiest to hear the songs of those creatures most like us, but Dionysius insists that the entire cosmos hums with divine presence and praise. To hear it requires the closest attunement of "intellection, and reason, and understanding, and touching, and sense perception, and opinion, and imagination, and name, and all other things"; it also requires recognizing that the divine in the cosmos exceeds all such knowing (On Divine Names 7.3). "God is both inaccessible and accessible to both sense and mind."[163]

Hierarchy—a word that Dionysius himself coined—may seem to be a sticking point for ecological readings of this theologian, as for Augustine (and not only for ecological readings, of course). Here Dionysius's thought is at its most complex. He offers the following definition: "Hierarchy is a sacred order (τάξις), an understanding (ἐπιστήμη), and an activity (ἐνέργεια) approximating as closely as possible to the divine" (On the Celestial Hierarchy 3.1). "Order" is the element that we most easily associate with the term *hierarchy*; it holds differentiation—the multiplicity and diversity of things—together with unity, rank, and harmony, resting on the Pauline metaphor of one body with many parts (1 Cor.12:12).[164] Hierarchy as order manifests on the spiritual, cosmic, and social planes; it occurs on scales both small and great. "Understanding" is the state of being mindful of such orderings, one might say. It is knowing oneself as both distinctive and relational, part of a larger whole, and it is unknowing oneself as either fully autonomous or strictly defined by one's place. It is knowing and unknowing oneself as a creature or made thing—given or called to being in the midst of an already teeming world. As Charles Stang frames the Dionysian insight, "We creatures, established in our place in the hierarchy, are offered the possibility of understanding—the unknowing/

knowledge of God—if we stand outside ourselves and heed the call of the creator."[165] To stand outside oneself—that is, to be in a state of ecstasy—is to be both here and elsewhere, at once self, other, and no-self, knowable and unknowable, understanding and ignorant. It is to be transported by divine desire or eros, the ecstatic love for god that is also from god. And it is this divine eros—alternately figured as light or Christ—that is the energy or "activity" flowing down and up and through all things, connecting, sustaining, enlivening, transforming.[166] Dionysian hierarchy is also synergy, then; it invites and enables a cooperation of energies and desires among a diversity of things.

But is hierarchy not also rigid and limiting and . . . well, *hierarchical*? No flat ontology or democracy of objects here. Some things are "over" and others "under"; some things and some names are better than others, are they not? Mary-Jane Rubenstein captures the interpretive dilemma well: "On the one hand, there is an incontrovertible order to the sacraments (and the clergy and the laity), mirroring the incontrovertible order of the cosmos. On the other hand, this cosmos refuses to stand still—circling, diving down, and spiraling in all directions. . . . On the one hand, the highest things most fully image God, while the lowest do so with limitations. On the other hand, the most 'inadequate and ridiculous' names are 'more suitable for lifting our minds up into the domain of the spiritual than similarities are.'"[167] There is rigidity when everything has its proper rank and place, yet that rigidity is undercut by the dynamism and fluidity of orderings that are (it seems) also ever emerging. There are superiority and inferiority when some are deemed more godlike than others, yet that superiority is undercut by Dionysius's ultimate insistence on the radical unknowability and otherness of god. If god is not comparable to *any* created thing, is the invocation of a ranking criterion of similarity not only ill-founded but potentially idolatrous? As Stang puts it, in a syntax that echoes Rubenstein's, "On the one hand, the entire structure of hierarchy is built upon the notion of relative likeness to God: some beings are more, and others less, like God. . . . On the other hand, the God to whom these beings are likened is also the unknown God, the God 'beyond being,' who is explicitly beyond any and all simile. . . . In a sense, hierarchy annuls itself."[168]

Thus it is that Dionysius moves hierarchically from the seemingly most appropriate names for god—good, existent, life, wisdom, power—to the seemingly least appropriate ones—angry, grieving, drunkard. The latter, tellingly, appear only in fragments cited from another lost or imaginary work (Mystical Theology 3): on the one hand, the dissimilar names are always already lost, erased, or unsaid; on the other hand, they expand in an endless, if not fully accessible, supplement. Dionysius recalls his own process of scriptural name-calling: "My argument traveled downward from the most exalted to the humblest categories, taking in on this downward path an ever-increasing number of ideas which multiplied with every stage of the descent." The pivot point between descent and ascent, procession and return, naming and negating, is not the apparent similarity or suitability of carefully culled names but the obvious dissimilarity or unsuitability of names that proliferate indiscriminately. Thus, notes Dionysius, scripture dubs god fire, water, ointment, or stone; lion, panther, leopard, or bear; and even worm, "the lowliest and most incongruous of all" (On the Celestial Hierarchy 2.5). Only by descending to the depths of linguistic inadequacy can Dionysius's thought soar beyond language: "My argument now rises from what is below up to the transcendent, and the more it climbs, the more language falters" (Mystical Theology 3). Negating "worm" or "drunkard," he can progress to negating even "the good." It is hard to let go of idols, but ultimately god is "beyond assertion and denial," as he asserts (5). Even negation must be negated. And it is in the negation of both affirmation and negation that "even hierarchy, determinacy, and certainty unsay themselves," as Rubenstein puts it.[169] A worm both is and is not god or a name of god. By the same token, the good both is and is not god, or a name of god. All things are not equal, but all things equally are-and-are-not god or names of god.[170]

To contemplate god is to tune into a cosmic superabundance. It is to listen to fragments of hymns salvaged from books whose sonorous plenitude can only be imagined. (It is also to look, touch, taste, and smell.)[171] The yearning of the creatures who sing and are sung—who carry and are carried by the music written in these books—strains to leap the gap not only between thing and the relational tangle of all-things but also between the knowable and nameable and that which exceeds apprehension, which cannot be called by name, which can only be hymned with silence. If this is the mystery of *things*, is it also *god*? Or as Rubenstein puts the question, "To what extent does God need to be thought of as 'beyond' or 'other than' a universe that is already non-identical and self-exceeding? Could we call that non-identical self-exceeding *itself* by the name of God?"[172] Dionysius insists that god is beyond being, that god is utterly alien to created things; at the same time, he presents creation itself as the incarnation of the god who is "being for whatever is" and "coming-to-be amid whatever happens" (On Divine Names 5.4). God is a name for the source and for the surplus of things, both in themselves and in their relations; it is also, then, a name for their potential resonances, for the love that circulates among them, for the cooperative possibilities that cannot yet even be imagined.

What does it mean to say that a stone or a tree or a worm can not only feel and know but also love? Anthropomorphism is a useful conceptual tool and also has its limits; it must ultimately be negated but so too must the negation be negated. A stone does and does not love, then. Put otherwise, it may be beyond human capacity to know what it is for a stone to experience "a unifying and co-mingling power that moves those higher to the care [πρόνοια] of those more needy, those of equal rank to mutual support, and finally those subordinate to a return to the stronger and the superior" (On Divine Names 4.15). And yet it may also be very important to acknowledge that a stone does experience something like that and also unlike that. It may be very important to pursue an "anthropology beyond the human," as Eduardo Kohn dubs it, thereby acquiring new conceptual tools "that can help us understand how we might better live in a world we share with other kinds of lives."[173] As Morton puts it, "If we have a future, we will have decided to look after all sentient beings. This decision is not calculating or utilitarian. At its limit, it is love."[174] And love is what pitches us beyond limits, including the limits of our own human-centered imagination—including, perhaps, the limits of our understanding of "sentience."

For both Augustine and Dionysius, divine love circulates through the cosmos, animating and connecting all things. What does this love desire? As fluid, it desires its own outpouring, in a flow of receiving and giving. As vision, it desires its own intensification, in the mutual beholding of all creatures. As sound, it desires its own amplification, in the resonant voices of all things. And so on. To love, as Dionysius puts it, is to belong "not to the self but to the beloved" (On Divine Names 4.13), and the beloved is all and beyond all. Sacrament and contemplation are the mediators of this more-than-all love. The erotic transformation envisioned thereby is communal and cosmic, perceptual and conceptual, bodily and spiritual, intimate and impersonal. It is painfully difficult—hard, hard work. And when it comes, it arrives as a gift.

Interlude: Words and Things

Words are fragments, open and inviting relation. Stamped into clay, carved into rock, inked onto paper, copied and recopied, printed and reprinted: they lure the eye, tickle the mind. Arriving as voice, they vibrate against ear, sonorous and ephemeral.

Words are footprints, but whose? We follow them across the page, tracking our prey. We give them voice: we let them animate our voices with their calling. We stutter, fall silent, drop the book: the *thing* so extravagantly exceeds the word that calls it.

Words invite relation, but they resist it too. The longer a word stares back at us—the longer it reverberates in the sound chamber of memory—the stranger it becomes, until it ceases to *mean* at all: at its limit, pure sensual *thing*. And yet we continue to write, to call, to read, to listen, to respond to the writing and calling of things, so long as we are able. We strain toward the place "where sound and sense mix together and resonate in each other, or through each other."[175] We long to experience once again the conjuring power of language.

Ontography: an odd word. The art of writing *things*. "Like a medieval bestiary, ontography can take the form of a compendium, a record of things juxtaposed to demonstrate their overlap and imply interaction through collocation."[176]

Compendium.

List.

Litany.

Incantation.

Poem.

Words "joined . . . by the gentle knot of the comma."[177]

"A repeated sorcerer's chant of the multitude of things that resist any unified empire."[178]

Imagine a book that archives *all* words, generates *all* magical lists and litanies.

Heaven, earth,
void, emptiness,
darkness, abyss,

breath, water,
light, darkness,
day, night,
morning, evening,
sky, sea, earth,
grass, seed, tree,
sun, moon, stars,
fish, birds,
reptiles, cattle,
humans, angels, demons,
oxen, lions, eagles,
panthers, leopards,
bears, worms.

The closer words get to things, the more they fall apart. Say them. Unsay them. Say them again. Listen to the clamor of voices!

In/Conclusion: Things, Practices, Piety

Ancient Christians seem to experience no sharp disjunction between the human and more-than-human world. They understand all things to be active, relational, and elusive—that is to say, alive in some sense or potentially so. Such an animist sensibility is not sustained only, or even primarily, through the teaching of theologians; indeed, Christian doctrine is often at odds with it. It is sustained rather through the cultivation of a range of embodied habits and feelings in which *things* feature as importantly as people do. In fact, ancient Christianity is in many respects less a set of beliefs than an array of practices that nurture a style of living reverently, mindful of the power inhering in all beings. This sensibility is what the ancients call *eusebeia* or *pietas*: *piety*, we may translate, however inadequately.

One of the most significant ways in which Christians cultivate relations with nonhuman things is through the cult of saints. Although apparently anthropocentric, the cult of saints, as we have seen, fosters a broad awareness of the "vitality of matter," as Jane Bennett names it.[179] In and as relics and icons, saints transcend their own humanity, to put it simply. At the same time, the elusive realm of the spiritual is made intimately accessible in the materiality of things; there the border between divine and human making, or nature and culture, begins to dissolve. Thus the shard of a human bone might be reclothed in the golden flesh of a reliquary. Alternately, the power and presence of a holy person might be transferred to a fragment of wood or rock subsequently adorned with metal or gems, to a trickle of oil caught in a flask, or to an image of the saint stamped in clay or painted on wood. Such a transformation of saint into thing—or rather, into multiple *things*, for relics and icons are eminently divisible and replicable—dramatically widens the range of the saint's relational capacities. The body-and spirit of the saint mingles with other materials, and new hybrids emerge. These hybrid things in turn create new relationships with other humans, as well as nonhumans, across many generations in some cases. Such relationships activate and engage the bodily senses and emotions of the humans involved and (we may imagine) of the nonhuman things too, through acts of mutual care, shared vulnerability, and riveting love.

Some relics and icons are small in scale, remaining mobile and manipulable and inviting intimate, even private relationships with their humans. Such holy things can also take on larger lives as parts of the assemblage of a church or a more extensive built complex. There relics might throb in the hidden heart of an altar, and images adorn walls or shine forth in mosaics arcing over an apse or carpeting a floor. Such images embody not only the figures of holy people but the narratives and scenes of the Bible writ large: typically, salvation history is presented on a cosmic scale. The buildings themselves have a vital force and relational energy of their own, moreover, shaping the flow of both space and time; they modulate the choreography of the liturgy while also mediating the relationships between interior and exterior, human craft and natural landscape. Holy things and places thus teach humans, among others, how to interact with the world as an open mesh of far-flung connection and intimate relation.

Words swirl around and through these things and places as well. They are inscribed on the walls and columns of churches. They echo in the readings, hymns, creeds, and sermons of the liturgy. In so doing, they speak to the material elements of the church buildings and of the wider cosmos, and those elements speak back. The image of *earth* on a floor mosaic in Petra, for example, reverberates with the first verse of the Bible, as that text is read and preached: "In the beginning the god created heaven and *earth*." And both image and word call to the *earth* who cradles the foundations of the church, offering her body for the stone and tile that in turn give the church body. In part through the mediation of words, holy things and buildings, as well as humans and other creatures, ultimately take their place in a cosmic scape of movement and song. For humans among others, nothing less is at stake than a radical transformation of sensation, emotion, and thought in the contemplation of creation.

Unlike *thing* or *object*, *creation* or *creature* implies a creator—a creator god, that is. The place of theism in ecological thought is highly contested, as we have seen. But so too is the place of piety. Throughout this book I have listened attentively to the voices of ecological thinkers who not only work outside the fields of religious studies and theology but who are deeply suspicious of both religion and theology as entrenched in dominative, human-centered, and other-worldly habits of thought and action. "A religious vocabulary is risky," when it comes to ecological thought, as Morton observes.[180] He also acknowledges, however, that there "might be seeds of future ways of being together in religion, as there are in art." The modes of coexistence that we need to learn are "almost unimaginable"; thus it is that they appear "as religion," he suggests.[181] Religion may offer not only habits of mind that widen our frame of reference and stretch our sense of possibility but also techniques for the crucial (if seemingly simple) work of cultivating attentiveness. "We shouldn't be afraid to withdraw and reflect."[182] Morton lauds the practice of contemplation, which (he insists) "implies an erotics of coexistence, not just letting things be."[183] I myself suspect that while theism may be dispensable for ecological thought and practice, some form of piety or cultivated reverence is not. And piety, as I understand it, is always in some sense performatively embodied. In its most mechanical aspect, it is carried by the momentum of repetition; yet it also gives rise to the vividness of novel encounter. Piety regulates relations between humans and other things, and it also disrupts and transforms those relations.

Christian cultures of piety are already at work in ecological thought, sometimes explicitly and sometimes implicitly, sometimes compliantly and sometimes subversively. As Douglas Christie has demonstrated with particular persuasiveness, those traditions may have more to teach, and more to learn as well, as they open themselves to the widest possible worlds of lively, lovable things.[184] I am intrigued that even stridently nontheological theorists of *things* not only turn to contemplation but also compose what they call litanies, referring to random lists of objects.[185] Uttered in a liturgical setting, such lists would constitute at once invocations and supplications, rendering objects present while also making a petition on their behalf; each pronouncement of a name would be met with a response, most traditionally "*Kyrie eleison*," or "Lord, have mercy." Language of lordship is admittedly problematic from an ecological (and a feminist) perspective, but a plea that mercy or compassion be bestowed on all things seems very much on target. Conveyed by the divine-human language of liturgy, that plea arises from all things and beyond, and is addressed to all things and beyond.

Bennett bends the language of "litany"—here apparently referring broadly to liturgical words chanted or recited collectively—in the direction of creed. Like other thing theorists, she composes her own—"a litany, a kind of Nicene Creed for would-be vital materialists." She does so, as she puts it, "at the end of [her] rope," having exhausted other conceptual and rhetorical formats. Her tone is lightly ironic yet nonetheless risks a certain earnestness, if not also a leap of faith:

> I believe in one matter-energy, the maker of things seen and unseen. I believe that this pluriverse is traversed by heterogeneities that are continually *doing things*. I believe it is wrong to deny vitality to nonhuman bodies, forces, and forms, and that a careful course of anthropomorphization can help reveal that vitality, even though it resists full translation and exceeds my comprehensive grasp. I believe that encounters with lively matter can chasten my fantasies of human mastery, highlight the common materiality of all that is, expose a wider distribution of agency, and reshape the self and its interests.[186]

Might we hear echoes of Dionysius's poetic theology in Bennett's creed—unintended, to be sure? The ancient theologian understands the divine as an enlivening, connective *energeia*, or activity, flowing through all things; things themselves are constantly receiving that *energeia* and passing it on. And while Dionysius provisionally endorses a hierarchical cosmic ordering based on relative superiority and inferiority, he ultimately chastens such a notion by insisting that no thing is more or less similar to the divine than any other thing: and if god ultimately remains unknowable, things do too. To understand this is to be on the path to salvation. But Bennett's creed is not only reminiscent of Dionysius's explicit theory. It also recalls the theory implicit in the practices of so many ancient Christians, who encounter a world full of vibrant yet elusive objects—holy, mysterious things inviting contemplation and evoking love.

If piety or reverence grounds the arts of coexistence, does it presuppose a god? We circle back to this question once again, and perhaps there is no end to the circling away and back. For ancient Christians, and not only ancient ones, god may name what evokes or draws the love of all things for all things—"the call that rises from things," as Chrétien has it, a call that is always also a response.[187] For others, it may no longer be possible to call upon the name of god at all, as Jean-Luc Nancy urges. "We must no longer seek either temples or deserts; we must abandon meditation."[188] The divine can no longer be encountered as such, he insists; it is "infinitely undone and scattered."[189] "In place of communion, in fact, there is the absence of the gods, and the exposure of each of us to the other: we are exposed to each other in the same way as we could, together, be exposed to the gods. It is the same mode of presence, without the presence of the gods. . . . Exposure takes place everywhere, in all places, for it is the exposure of all and of each, in his solitude, to not being alone."[190] Here we may be reminded of Augustine's vision of the end times, when "god will be seen by each one of us in each one of us" (City of God 22.29): contemplation of god becomes the exposure of each to all the others. For all its differences, Nancy's radical a-theology is here partly continuous with traditions of both negative theology and the theology of things implicit in late ancient Christian piety. It is also profoundly ecological. Eschewing the "sacred enclosures" of the god, Nancy seeks "other tracks, other ways, other places for all who are there."[191]

Like a hot, dry desert wind scouring the ruins of ancient churches long since exposed to the elements, ecological thought opens both theology and piety to their widest possibilities. Perhaps, at their best, they were always already seeking those possibilities.

Postlude: The Things That Matter

What is a holy thing?

What is *not* a holy thing?

Having spent the day tramping around archaeological sites, I sit with my friend and our guide in a Jordanian cave that has been fashioned into a space of hospitality. Benches are covered in tapestries and pillows, and the floor is carpeted; the rock-hewn walls are covered in . . . well, all kinds of stuff, from business cards, to photos of the royal family, to brightly colored homemade textiles. Two long tables are covered in things as well, artfully organized by kind—ancient coins, bits of pottery, pieces of oddly shaped stone or wood, some remarkable necklaces made from iridescent insect wings. Our host was not expecting to entertain us any more than we were expecting to be his guests, but as far as I can tell, this only enhances the pleasure of the visit for all of us. He serves us tea and eagerly shows us the objects displayed on the walls and tables, caressing the items, naming them, picking them up and rearranging them, telling their stories in a language we do not speak. In some cases, he wants us to photograph ourselves with him standing next to a particular item. In other cases, he makes a gift of a thing—a bead, a polished stone fashioned as a pendant. He accepts a modest gift—a business card to add to his collection. I am coughing, and he insists that I snort a concoction that he has on hand, giving me some dry herbs to take with me so that I can make my own tincture later. Thrusting a photo album into my lap, he eagerly turns the pages, touching the snapshots and pointing out highlights. Most depict him posed with a person I am expected to recognize; I smile broadly and try to look appropriately impressed, covering my embarrassed ignorance. As we leave, he pulls us into his garden, a tiny patch of ground next to the road on which several small fruit trees crowd; some we can identify, others not. He introduces us to each tree in turn, describing its properties and uses, urging us to look, smell, touch, taste. We receive more gifts, familiar apricots and some small, hard, green fruits that, as he indicates with winks and nudges, are thought to have aphrodisiac powers.

Even in the moment, I am vaguely aware that the things I have been given will have lost much of their aphroditic magic by the time I have emptied my pockets and unpacked my backpack in the hotel. The thought is tinged with sadness and—more surprisingly—a sense of loneliness.

Perhaps I write these words to restore the holiness of things for myself, to recall them to their special glimmer and sweetness, to recapture their abiding allure. I write to remember that I am not alone, that I am (on the contrary) impossibly *exposed*, exposed to all the other things *as if to gods*. I write to recover lost arts and erotics of coexistence.

EPILOGUE
Worm Stories

"Worms have played a more important part in the history of the world than most persons would at first suppose": thus writes Charles Darwin in the concluding chapter of his final monograph, entitled *The Formation of Vegetable Mould Through the Action of Worms, with Observations on Their Habits.*[1] Marveling at the vast numbers, muscular strength, tireless diligence, and surprising intelligence of earthworms, Darwin observes that in England "the whole superficial bed of vegetable mould passes through their bodies in the course of every few years." In other words, the topsoil required for the flourishing of so much plant life is the product of worms' collective digestive processes. What worms cannot consume and metabolize, they cover over with their excrement. As Darwin explains, "Archaeologists ought to be grateful to worms, as they protect and preserve for an indefinitely long period every object, not liable to decay, which is dropped on the surface of the land, by burying it beneath their castings." Archaeologists might also have cause to feel grumpy, as it is likewise due to the constant undermining of earthworms that "many monoliths and some old walls" have collapsed, as Darwin also notes. Moreover, by bringing fine grains of soil to the surface, worms are active in the processes of erosion that constantly change the contours of the planet. On the closing page of his book, Darwin reiterates, "It may be doubted whether there are many other animals which have played so important a part in the history of the world, as have these lowly organized creatures."[2]

There is both an implicit politics and an implicit theology at work in Darwin's praise of worms. With his "worm stories," the famous naturalist gently but persistently questions hierarchies of both class and species. Darwin's worms humble the human and call attention to the interdependence of all creatures. In our own time, they continue to inspire radically inclusive democratic theories in which "the appropriate unit of analysis . . . is neither the individual human nor an exclusively human collective," as political theorist Jane Bennett puts it.[3] "Even if a convincing case is made for worms as active members of, say, the ecosystem of a rainforest, can worms be considered members of a *public*? What is the difference between an ecosystem and a political system?" she asks.[4] At the same time, Darwin's worms replace Christian concepts of creation and resurrection with what psychoanalyst Adam Phillips dubs "a secular maintenance myth." I myself might call it a "worm theology." "It is as though the earth is reborn again and again, passing through the bodies of worms," as Phillips puts it.[5] For Darwin, what mattered was "not life after death . . . but life with death," notes Phillips. Appropriating the theological idiom, he declares, in the spirit of Darwin, "When transience is not merely an occasion for mourning, we will have inherited the earth."[6]

Worms have appeared spontaneously at several points in this book, as I have attempted to articulate a three-faceted ecopoetics, engaging the philosophical, literary, and material aspects of late ancient Christianity. Admittedly, their appearances have sometimes been modest; some readers may scarcely have noticed them. Let us return to these worms, then, and meditate on them just a bit longer here at the end. We might think of them, like Darwin's worms, as creatures easily overlooked yet nonetheless playing a significant historical role—in this case, shaping not the contours of the earth's surface but the contours of the Christian theological imagination. As we shall see, worms enter Christian thought largely through biblical passages—passages that are continuously renewed by the exegetical creativity of their interpreters. From there, they open new paths of thought. Collectively, they may serve as a closing reminder that the same Christians who seemingly had their eyes fixed steadily on the invisible heavens above also never quite lost sight of the loamy earth under their feet.

Thinking Cosmologically with Worms: Afterlife or Life Now?

In ancient Jewish texts, the worm—here, as so frequently, not an earthworm but an insect larva, or maggot—typically serves as a figure for human mortality, corruptibility, and transience. Frequently, the worm in this guise issues a call to humility on the part of humans, as in this verse from the Wisdom of Ben Sirach: "Be as humble as possible, for the hope of humanity (*tik-*

vat enosh) is a worm (*rimah*)" (Sir. 7:17, from the Hebrew). The Mishnah repeats the sentiment: "Rabbi Levitas of Yavneh used to say: Be exceedingly humble of spirit, for the hope of humanity (*tikvat adam*) is a worm (*rimah*)" (Mishnah Avot 4:4). That is to say, ultimately, humans like other living beings will die and disintegrate, returning to the earth with a little help from the digestive labor of worms. That is our only realistic expectation, but maybe it is also, indeed, our best hope—current embalming practices notwithstanding. Returning as dust to dust, humans may live on through the lives of other creatures.

When does another kind of afterlife enter the history of Jewish and Christian thought, and what happens to our worms when it does? Hindsight may offer misleading clarity: contemporary readers are likely to detect other-worldliness where it was not originally present. The biblical book of Mark reads as follows: "And if your eye causes you to stumble, tear it out; it is better for you to enter the kingdom of God with one eye than to have two eyes and to be thrown into Gehenna, *where their worm never dies, and the fire is never quenched*" (Mark 9:47–48). Here the gospel writer alludes to the ominous last verse of Isaiah, where God speaks through the voice of the prophet: "And they shall go out and look at the dead bodies of the people who have rebelled against me; for *their worm shall not die, their fire shall not be quenched*, and they shall be an abhorrence to all flesh" (Isa. 66:24). In this passage, Isaiah describes bodies displayed in a suspended state of humiliation, corrupting and burning simultaneously, without being dissolved—not an afterlife, it would seem, so much as an instructively horrifying shaming of corpses. As for the gospel writer, he may refer to a fiery garbage dump in the environs of Jerusalem as the resting place of disgraced bodies, though there is much scholarly debate on this point. However we understand the reference to Gehenna, it is significant that the gospel of Mark is not the only Second Temple text to pick up on Isaiah's theme of divine punishment by worm and fire. In the apocryphal book of Judith, the heroine cries out, "Woe to the nations that rise up against my people! The Lord Almighty will take vengeance on them in the day of judgment; he will send *fire and worms* into their flesh; they shall weep in pain forever" (Jth. 16:17). In fact, the passage from the Wisdom of Ben Sirach quoted above—"for the hope of humanity is a worm"—is rendered in the Septuagint (and in most other subsequent Christian translations) as "for the punishment of the ungodly is *fire and worms*" (Sir. 7:17, LXX). The worm has thus been transmuted from a figure of human mortality into a figure of divine vengeance. It comes to signify not the mutual enfolding of life and death but the promise of eternal punishment after death. In other words, the first steps have been taken toward consigning the worm to hell. It seems that the worm must be banished if life is to triumph over death and this world is to be traded in for another, more perfect one.

This particular worm story gets more interesting when the controversial doctrine of bodily resurrection is further developed by later Christian theologians—Origen of Alexandria on the one hand and Augustine of Hippo on the other. Each of them has to defend the doctrine in the face of sharp criticism from Platonic philosophers as well as other Christians. As Enrica Ruaro has shown, worms run like a leitmotif through the philosopher Celsus's attack on the Christian doctrine, to which Origen responds.[7] Celsus condemns Jews and Christians alike as "a cluster of bats or ants coming out of a nest, or frogs holding council round a marsh, or worms assembling in a muddy corner, disagreeing with one another about which of them are the worst sinners." Christians, he continues, are like worms who absurdly believe that they are the most godlike of all creatures, ranking even higher than the stars (Origen, Against Celsus 4.23). But the core issue for Celsus is this: since there is no difference between the body of "a bat or a worm or a frog or a human," all being equally corruptible (4.52), what use does a human soul have for a body after death? "This is simply the hope of worms," Celsus concludes (5.14), in an odd (and apparently accidental) echo of the Hebrew text of Ben Sirach. Celsus mocks those who identify so closely with their bodies that they want to hold onto them forever: such humans might as well be worms.

Yet for Celsus, resurrection is the hope of worms in another sense as well. In antiquity, worms were thought to arise directly from mud, wood, or animal corpses; they had an unusual mode of generation, in other words. Celsus quotes "popular opinion" that "a snake is formed out of a dead man, originating from the marrow of the spine, and a bee from an ox, and a wasp from a horse, and a beetle from an ass, and in general worms from most animals" (4.57). Thus worms can signify for Celsus both the corruptibility and the regenerative capacity of the human body.

In responding to Celsus, Origen predictably rejects the notion that Christians are like worms in their love of the flesh, insisting that resurrected bodies will be refined spiritual bodies, qualitatively different from the grossly material ones that souls currently inhabit. However, as we have seen, he does, somewhat surprisingly, seem to accept the notion that the process by which worms are generated from other matter may be a fit metaphor for the generation of the resurrected body from the corruptible one. What Celsus has missed is that the seminal Word of God is at work in that transformative process (4.57). Elsewhere, Origen observes that Christ refers to himself as a worm (Commentary on Luke 14.8). This is an argument that arises from the widespread assumption among early Christians that Psalm 22 in its entirety represents Jesus's last words spoken from the cross.[8] Both Mark and Matthew depict Jesus as citing the first verse of the psalm: "My God, my God, why have you forsaken me?" (Mark 15:34, Matt. 27:36). The sixth verse of the psalm reads, "I am a worm and not a human, scorned by others, and despised by the people." Origen suggests that Christ, in saying "I am a worm," references his own spontaneous generation from the womb of a virgin. Thus a passage that can also be read as protest against humiliation on the cross is transformed into a positive reference to Christ's miraculous birth—or, alternately, to his own resurrection, as Damasus of Rome and Isidore of Seville both have it.[9]

If Origen is able to appropriate the worm denigrated by Celsus as a positive figure for virginal birth and bodily resurrection—the worm of heaven, we might say—he does not show the same interest in the worm of hell that other ancient commentators discovered in biblical passages referring to punishment by "fire and worm." Famously, Origen taught that all creatures would, in the end, be restored to their original perfection, an optimistic position that leaves no room for a doctrine of eternal punishment. He does refer to the process of purifying the soul as a kind of refining fire that burns away sins, but he interprets that fire metaphorically and psychologically: "Every sinner lights the flame of his own fire himself, and is not plunged into some fire that has been lit before that moment or exists in front of him. . . . The conscience is agitated and pierced by its own pricks" (On First Principles 2.10.4).

Augustine is considerably less optimistic than Origen, whose views he explicitly dismisses in the twenty-first book of his *City of God*. There he writes, "I must now, I see, enter the lists of amicable controversy with those tenderhearted Christians who decline to believe that any . . . shall suffer eternally, and who suppose that they shall be delivered after a fixed term of punishment. . . . In respect of this matter, Origen was even more indulgent; for he believed that even the devil himself and his angels . . . should be delivered from their torments. . . . But the Church, not without reason, condemned him for this and other errors" (City of God 21.17). In the course of defending the theory of eternal punishment of the damned, Augustine cites the "fire and worm" biblical passages mentioned earlier. But here too he must confront the views of those who, like Origen, interpret not only fire but also worm metaphorically or psychologically. In the case of the worm, his interlocutors invoke the support of a passage from Proverbs: "Like a moth in clothing or a worm in wood, sorrow gnaws at the human heart" (Prov. 25:20). Just as fire can be read as a sign for the burning pain of a soul in the state of sin, so the worm can be read as a sign for the gnawing agony of the soul's consequent anguish. Augustine's own position is that eternal punishment must involve bodily and spiritual suffering, though he still grants his readers some latitude: "Let each one make his or her own choice, either assigning the fire to the body and the worm to the soul—the one figuratively, the other really—or assigning both really to the body" (City of God 21.9). Thus the worm of hell (unlike the fire of hell) may be taken either literally or metaphorically.

The worm also appears in one other guise in Augustine's discussion of eternal punishment in the *City of God*, namely, as an analogue to the human facing punishment. "In springs of water so hot that no one can put his or her hand in it without harm, a species of worm is found, which not only lives there, but cannot live elsewhere," Augustine tell us; just so, human bodies can burn eternally without being consumed, he concludes (21.2). Thus the theologian's attempt to make eternal punishment credible leads him to invoke the nearly incredible marvels of mortal creatures adapted to a wide range of landscapes—or in this case, waterscapes. Intriguingly, similar worms have been a source of wonder more recently. In 1977, scientists exploring deep-sea hot springs in the Galapagos Rift discovered "communities of foot-long clams, white crabs, and giant white tube worms with bright red plumes waving in the current—all species that had never been seen before." The worms, living in near-boiling water, grow up to six feet long and house colonies of symbiotic bacteria that aid their hosts in transforming hydrogen sulfide absorbed from the hot springs into carbohydrates and proteins through a process similar to photosynthesis. This discovery "did nothing less than change our understanding of life on the planet," scientists proclaimed.[10]

It seems, then, that worms continue to be good to think with cosmologically. Historically, Christian exegetes have discovered figures of painful disintegration, transformative rebirth, and remarkable adaptation—rather Darwinian themes, in fact. Even depictions of the damned being punished by the undying Markan worm send an ambiguous message, as well they might: eternal corruption is a deeply paradoxical concept. A medieval fresco from the Serbian church in the monastery at Dečani shows a tight cluster of nine beardless youths clad in loincloths, their lean, well-articulated torsos overlaid with white squiggles that drift upward from the gently mounded foreground (Figure 14). These worms are aesthetically pleasing, even decorative, and the young men whose languid bodies they traverse appear somber but not disgusted, alarmed, or even pained, as they lean into one another intimately.

Figure 14. "Worms That Sleepeth Not." Fresco, fourteenth century, Dečani Monastery Church. Photographer, Ivan Drpic.

Ignoring the viewer, they gaze steadily into each others' faces. The figure on the lower right rests his head on his own shoulder almost seductively. Perhaps we are meant to assume that the youths are contemplating their sins, but it is also possible to imagine them expressing a serene acceptance of the mortality that haunts the vitality of even their youthful, erotically charged bodies. From the perspective of eternity, living and loving and dying are not sequential but simultaneous processes, mutually enfolded one in the other.

Is it possible that these wriggling biblical worms, having been banished to the afterlife, have subtly undermined the very "walls" that were intended to keep "the history of the world" separate from eternity and hell separate from heaven? Burrowing across barriers, they may have smuggled in something of bodily life, in its malleable resilience and poignant vulnerability—"prone to many deaths, and shadowed by the reality of death,"[11] and yet also pulsing with a remarkable liveliness.

Thinking Narratively with Worms: Autonomy or Mutual Dependence?

One set of worm stories that readers are likely to remember well are the stories about Simeon, the first and most famous of the so-called stylite saints, who observed the distinctive ascetic practice of living on small, open platforms atop pillars. Simeon's worms seem to emanate from different biblical quarters, namely, the book of Job. The initial story takes place before he starts building pillars, at a point when he is still attempting to fit into communal life in a monastery—attempting, but not succeeding, and worms are a big part of the problem, as we recall. It starts when Simeon steals a rope and wraps it tightly around his body, hiding his deed of mortification with a rough tunic. A year later, the flesh under the rope is putrid. As his biographer Antonius reports, "Because of his stench no one could stand near him, but no one knew his secret. His bed was covered with worms, but no one knew what had taken place" (Life of Simeon 5). Finally, one of the monks goes to the abbot, complaining, "His bed is full of worms, and we simply cannot bear it" (6). Astonished, the abbot goes to inspect and finds that Simeon's bed is indeed "full of worms" (7). After dubbing Simeon a "new Job," he orders the monk stripped, which proves quite a difficult task. The reader is told, "there was no guessing how many worms were on him" (8). The abbot's comparison of Simeon to Job is apt, yet Simeon seems determined to outdo even the biblical figure by embracing the very fate that Job laments. "My flesh is clothed with worms and dirt," cries Job (Job 7:5). "Where then is my hope?" he asks (Job 17:15). *Your hope is the worm*, Simeon might have answered.

Much later, when Simeon has mounted his final and highest pillar, he develops a tumor on his thigh, "just as happened to the blessed Job," as Antonius reports, so that he has to stand on one foot for two years. As Simeon remains balanced with yogic virtuosity in a kind of "tree" pose, his thigh becomes so massively infected by maggots that the wriggling creatures fall like rain to the earth sixty feet below. The saint's followers must spend all of their waking hours retrieving the worms and carrying them back up the ladder to replace them on Simeon's body, where he encourages the worms to feed from his flesh (Life of Simeon 17). We may be reminded of the Dečani fresco of the young men who are so seemingly serene as the wriggling white worms cover their skin or of the more recent performance by Gina Pane, in which the artist, her face covered with live maggots, embraces "two fleshes living together, one nourishing itself from the other: process of life in a continuum of time," as she puts it.[12] In this context, Simeon and his worms may suggest less the disintegration of a human corpse than a profound symbiosis of living creatures.

Simeon's hope is the worm because it affords him the experience of radical intimacy with some of the "strange strangers" with whom humans share the planet and sometimes the space of their own bodies. As it happens, the worm may also have been his hope in another sense. That is, Simeon—should we choose to read this text as a kind of history—may have been the beneficiary of so-called larval therapy, whereby larvae or maggots effectively remove dead, damaged, or infected tissue to improve the healing potential of the remaining healthy tissue, while also producing antiobiotic-like agents and inhibiting bacterial growth through secreted ammonia.[13] From this perspective, Simeon is not only living sacrificially for the sake of hospitality but also opening himself to a kind of adaptive coexistence that serves both him and the worms very well.

Thinking Liturgically with Worms: Hierarchy or the Democracy of All Creatures?

The final set of worm texts that I want to consider relate to rhetorical and liturgical performances of praise of creation. The first comes again from Augustine:

> We must admit that a weeping human is better than a happy worm. And yet (*tamen*) I could speak at great length without any falsehood in praise of the worm (*vermiculus*). I could point out the brightness of its coloring, the slender rounded shape of its body, the fitness of its parts from front to rear, and their effort to preserve unity as far as is possible in so lowly a creature. There is nothing anywhere about it that does not correspond to something else that matches. What am I to say about its soul animating its tiny body? Even a worm's soul causes it to move with precision, to see things suitable for it, to avoid or overcome difficulties as far as possible. Having regard always to the sense of safety, its soul hints much more clearly than its body at the unity which is the foundress (*conditrix*) of all natures. I am speaking of any kind of living worm. (On True Religion 41.77)

Here Augustine upholds a hierarchy of creatures in which the human is far above the lowly worm—and this is not just a worm (*vermis*) but a *little* worm (*vermiculus*). However, there follows a telling "and yet" (*tamen*). It is just a little worm. *And yet* Augustine, ever the rhetorician, can deliver a speech in its praise. He could go on forever, if given half a chance, he suggests: one thinks of Darwin's more than three hundred pages. Augustine's speech proclaims that the worm is just exactly as it should be. Moreover, its shimmering vitality hints at its crucial place in the greater totality of "natures" that gives rise to it. By thus drawing our attention so effectively not only to the worm's worthiness but also to its relation to the rest of creation, Augustine at least partly undercuts the very hierarchy that he inscribes at the outset. When it comes to the beauty of the universe, what matters, as he puts it, is not the parts but the whole: it is the whole that is perfect, "whether its beauty is seen at rest or in movement" (40.76). How then can any one part, whether at rest or in movement, be judged as greater or lesser?

As we have seen, a similar question can be posed with regard to the thought of the later theologian known by the pseudonym of Dionysius. Like Augustine, but even more explicitly, Dionysius describes creation in terms of hierarchy—a word that he himself apparently coins. This hierarchy is based on the principle of relative likeness: some things are understood to be more like the divine and some less so, although the hierarchy joins them all in a web of relationality, as Augustine's theory of beauty also suggests. At the same time, Dionysius asserts quite radically that the creator god is utterly unknowable and beyond all comparison to any created being. And if the divine is beyond all likeness, a hierarchical order based on relative likeness cannot ultimately be sustained. Is not every creature equally dissimilar to the alien god, equally inappropriate to reflect the divine image, and by the same token equally similar and equally appropriate? In order to make this point, and thereby to guard against an idolatry of particular "godlike" things, Dionysius undertakes a strategy of privileging precisely those figures that are, according to the hierarchy, the least like the god. In this context he comments, "I will add what seems to be more dishonorable than all, and the most incongruous, that is, that distinguished theologians have shown [the god] to us as representing itself under the form of a worm" (On the Celestial Hierarchy 2.5). As Ruaro points out, Dionysius here surely refers to Psalm 22—"But I am a worm and no human"—when he states that the god represents itself as a worm, given that this speech is traditionally attributed to Christ, as we have seen, and here by extension to the godhead itself.[14] For Dionysius, this is just one more instance of scripture singing god's praises by calling the name of every creature, from the seemingly "highest" to the seemingly "lowest." The universe itself is a hymn to the divine, expressing simultaneously the desire of every creature for god and of god for every creature, worm as much as human.

Ultimately, Dionysius's liturgical theology of the absurd (a declaration that god is like a worm), like Augustine's performative rhetoric of the absurd (a speech in praise of the little worm), may be suspected of preserving hierarchy, insofar as the worm must remain lowly if it is to function in its absurdity. Yet as we have seen, it is also the case that hierarchy as a reflection of relative godlikeness ultimately undoes itself: no one creature can claim to be more or less like the god than any other. Perhaps we should say, then, not that a worm is less worthy than a human but that the worm stands as witness for all those beings who escape the notice of most humans most of the time. When humans do take note of them, they are inclined to undervalue them or even to be repelled: from an ordinary human perspective, worms are humble creatures indeed. In contrast, in both a Darwinian and a theological context, it is worms who humble humans. Worms remind humans of how little we know of our place in this ever-shifting and intricately relational universe, and (for those theistically inclined) of how little we know of the god of which this ever-shifting and intricately relational universe both is and is not an image.

The Hope of Worms: Passages to a Future

Dark cosmologies. Queer saints. Attunement to the liveliness of things. These are the remains of an ancient ecopoetics. They may seem strange—not quite what we expected, perhaps. At the same time, it is as if we were already waiting for them to arrive. It is as if they were already waiting for us too, at the edge of an open and emerging future.

Leaning into an open future means allowing history to open up as well. If we cannot restore what once was, we had best learn to live as well as possible with what might be—a first step. In so doing, we may come to realize that history is nothing but open moments—every "once was" has always been full of "might be's." Thus we tell our stories only to tell them again, otherwise. Better yet, our stories tell us; they flow through us, shifting with the tides of time. The mesh of interconnection is not only spatial but also temporal. These ancient Christian ideas, persons, and things are far away and long ago, and they are also among our intimate relations. We have made them so by attending to them. Perhaps they were always already attending to us as well.

Dark cosmologies. The realm of becoming—"the all," as the Greeks named it—is mysterious, never fully in our grasp. It hovers between the nothing and something of intelligibility, the absence and presence of sensation, evoking horror and wonder at once. Its forms—call them creatures, if you will—are ever emerging and dissolving; they exist only in relation to one another and to the all itself. They are constrained by all those others, even as each is impelled by its manifold pasts, lured by its manifold futures. They are also each free, to some significant degree: we cannot know whither their desires will lead them, how their received possibilities will materialize; we cannot know where this *all* is heading. But we can and do know that every creature matters: we are in it together. Our hope as humans is the hope of worms too.

Queer saints. Saints are luminous beings who draw our attention and inspire our awe. They break through the self-constraining limits of humanity by transgressing the boundaries that separate humans from other animals, including such strange animals as angels and demons. Thus saints are no longer *human* beings, or not exclusively so. Eschewing the straight and narrow of scripted roles and life trajectories, they live at the edges of their own bodily and psychic capacities—abject vulnerabilities, startling strengths. What might yet be possible for such queer creatures, who have merged into the landscape of their more-than-human entanglements, who have opened themselves to the hauntings of their own horrors—who have allowed their flesh to be traversed by the flesh of worms?

Attunement to the liveliness of things. The power of sanctity exceeds the human, exceeds even the animal: it seeps and spreads and sticks to all manner of things. Bodily remains; cloth and oil and wood; molded, carved, or painted images of the holy ones—all of these objects, and more, possess liveliness and force. In the face of the very particular and distinctive potencies of individual beings, categories do not always hold. A worm dropping from a saint's body may be called a pearl, blurring the line between animal and thing: is that a demotion or a promotion for the worm? A pearl, once part of a hybrid assemblage of shell, mollusk, and grit, may come to adorn the gilded housing of a bit of bone and dried blood or a shard of wood, announcing the still greater preciousness of such humble fragments. For those who cultivate intimate relationships with apparently inanimate things, conventional hierarchies of being are topsy-turvy, and the line between the living and the nonliving begins to dissolve. Things are animate too, in short. Things are relational as well. The line between nature and art also begins to dissolve. Things are not static but, like animals, ever emerging, sometimes in conjunction with human agency, as in the case of a reliquary or a mosaic, sometimes in conjunction with other kinds of agents, as in the case of a pearl or topsoil. Each thing is irreducibly itself and can also only be thought in relation to all things. And what, finally, *is not* a thing, when a worm is a pearl and a human is a bone and a god's power and presence saturates a splinter of wood? We are back to the all, then. To *all things*.

Creation is the god singing all things, in an outpouring of love. It is also all things singing the god, through their own being, in an outpouring of love in return. It is all things singing the love of all things. Such is the hope of humans, as of worms. Sometimes the song is unbearably sorrowful. Sometimes it falters, stopping altogether. Sometimes it soars with joy. Can you hear it?

Notes

Introduction

Note to epigraph: Timothy Morton, *The Ecological Thought* (Cambridge, MA: Harvard University Press, 2010), 134.

1. Morton, *Ecological Thought*, 19.

2. This is a complicated "we," as neither the responsibility for ecological degradation nor its effects fall evenly on the human species; see Andreas Malm and Alf Hornborg, "The Geology of Mankind? A Critique of the Anthropocene Narrative," *Anthropocene Review* 1 (2014): 62–69. For a defense of the concept of the anthropocene and the significance—and challenge—of thinking the human on the scale of the species, see Timothy Morton, *Dark Ecology: For a Logic of Future Coexistence* (New York: Columbia University Press, 2016), 14–25.

3. Christopher Schliephake, "Introduction," in *Ecocriticism, Ecology, and the Cultures of Antiquity*, ed. Christopher Schliephake (Lanham, MD: Lexington, 2017), 4.

4. J. Donald Hughes, *Pan's Travail: Environmental Problems of the Ancient Greeks and Romans* (Baltimore: Johns Hopkins University Press, 1994), 196.

5. Bill McKibben, "Foreword," in *Toward an Ecology of Transfiguration: Orthodox Christian Perspectives on Environment, Nature, and Creation*, ed. John Chryssavgis and Bruce V. Foltz (New York: Fordham University Press, 2013), xiii.

6. Todd LeVasseur and Anna Peterson, "Introduction," in *Religion and Ecological Crisis: The "Lynn White Thesis" at Fifty*, ed. Todd LeVasseur and Anna Peterson (New York: Routledge, 2017), 2.

7. Lynn White Jr., "The Historical Roots of Our Ecological Crisis," *Science* 155, no. 3767 (1967): 1205.

8. White, "Historical Roots," 1207.

9. Willis Jenkins, "After Lynn White: Religious Ethics and Environmental Problems," *Journal of Religious Ethics* 37 (2009): 288.

10. Donovan O. Schaefer, *Religious Affects: Animality, Evolution, and Power* (Durham, NC: Duke University Press, 2015), 9.

11. Morton, *Dark Ecology*, 131.

12. See Christopher Key Chapple, "Lynn White Jr. and India: Romance? Reality?" in LeVasseur and Peterson, *Religion and Ecological Crisis*, 110–20.

13. Douglas E. Christie, *The Blue Sapphire of the Mind: Notes for a Contemplative Ecology* (New York: Oxford University Press, 2013).

14. See the Forum on Religion and Ecology at Yale (website), School of Forestry and Environmental Studies, Yale University, accessed July 28, 2017, http://fore.yale.edu/.

15. Morton, *Ecological Thought*, 134. Morton is especially drawn to the resources of Buddhism for thinking ecologically; see Marcus Boon, Eric Cazdyn, and Timothy Morton, *Nothing: Three Inquiries in Buddhism* (Chicago: University of Chicago Press, 2015).

16. Whitney A. Bauman, "What's Left (Out) of the Lynn White Narrative?" in LeVasseur and Peterson, *Religion and Ecological Crisis*, 167.

17. See for example the defense of Protestant Christianity by Michael S. Northcott, "Lynn White Jr. Right and Wrong: The Anti-Ecological Character of Latin Christianity and the Pro-Ecological Turn of Protestantism," in LeVasseur and Peterson, *Religion and Ecological Crisis*, 61–74.

18. While White has frequently been viewed as "anti-Christian," Matthew T. Riley brings out his positive commitment to the reform of Christian theology ("A Spiritual Democracy of All God's Creatures: Ecotheology and the Animals of Lynn White Jr," in *Divinanimality: Animal Theory, Creaturely Theology*, ed. Stephen D. Moore [New York: Fordham University Press, 2014], 241–60).

19. Scott Knickerbocker, *Ecopoetics: The Language of Nature, the Nature of Language* (Amherst: University of Massachusetts Press, 2012), 17–18.

20. David L. Miller, "Theopoetry or Theopoetics?" *Crosscurrents* 60, no. 1 (March 2010): 8.

21. Angela Hume, "Imagining Ecopoetics: An Interview with Robert Hass, Brenda Hillman, Evelyn Reilly, and Jonathan Skinner," *Interdisciplinary Studies in Literature and Environment* 19, no. 4 (2012): 756. For a concise discussion of how one might define "ecopoetics," see also Jonathan Skinner, "Ecopoetics," *Jacket2* (2011), accessed December 11, 2017, http://jacket2.org/commentary/jonathan-skinner.

22. Hume, "Imagining Ecopoetics," 761.

23. John Sallis, *Platonic Legacies*, SUNY Series in Contemporary Continental Philosophy (Albany: State University of New York Press, 2004), 93.

24. See especially Timothy Morton, "Queer Ecology," *Proceedings of the Modern Language Association* 125, no. 2 (2010): 273–82.

25. See especially Jane Bennett, *Vibrant Matter: A Political Ecology of Things* (Durham, NC: Duke University Press, 2010).

1. Beginning Again with *Khora*: Traces of a Dark Cosmology

Note to epigraph: John Sallis, *Platonic Legacies*, 93.

1. Note that my own habit is to transliterate χώρα as *khora*, but I follow Sallis in an alternate transliteration when invoking his distinctive use of the term *chorology*.

2. See, e.g., Julia Kristeva, "Revolution in Poetic Language," in *The Kristeva Reader*, ed. Toril Moi (Oxford: Basil Blackwell, 1986), 89–136; Kristeva, *Powers of Horror: An Essay on Abjection*, trans. Leon S. Roudiez (New York: Columbia University Press, 1982); Luce Irigaray, *Speculum of the Other Woman*, trans. Gillian C. Gill (Ithaca, NY: Cornell University Press, 1985; French ed., 1974), esp. 168–79; Irigaray, *An Ethics of Sexual Difference*, trans. Carolyn Burke and Gillian C. Gill (Ithaca, NY: Cornell University Press, 1993), 34–55; Jacques Derrida, "How to Avoid Speaking: Denials," in *Derrida and Negative Theology*, ed. Harold Coward and Toby Foshay (Albany: State University of New York Press, 1992), 73–142; Derrida, "Khōra," in *On the Name*, trans. David Wood, John P. Leavey, and Ian McLeod (Stanford, CA: Stanford University Press, 1995), 87–127.

3. For an exception to this general rule, see Rebekah Sheldon, "Form/Matter/Chora: Object-Oriented Ontology and Feminist New Materialism," in *The Nonhuman Turn*, ed. Richard Grusin (Minneapolis: University of Minnesota Press, 2015), 193–222.

4. See also Melissa Lane, *Eco-Republic: What the Ancients Can Teach Us About Ethics, Virtue, and Sustainable Living* (Princeton, NJ: Princeton University Press, 2012). Lane argues that Plato's Republic is "a primer in the functioning of political possibility" (22) that may offer a paradigm for the transformations of soul and city required to meet the challenges of sustainability.

5. For Plato's *Timaeus* and *Critias*, I have used the Greek edition in Plato, *Timaeus, Critias, Cleitophon, Menexenus, Epistles*, trans. R. G. Bury, Loeb Classical Library 234 (Cambridge, MA: Harvard University Press, 1929). Unless otherwise noted, translations are my own.

6. See Derrida, "Khōra," 89: "*Khōra* reaches us, and as the name."

7. Morton, *Ecological Thought*, 16. See also Morton, *Dark Ecology*.

8. John Sallis, *Chorology: On Beginning in Plato's* Timaeus (Bloomington: Indiana University Press, 1999), 5. My reading of *khora* is heavily indebted to Sallis throughout.

9. Sallis, *Platonic Legacies*, 4.

10. Sallis, *Chorology*, 49.

11. See Sallis, *Chorology*, 50.

12. See Sallis, *Chorology*, 52. The question of how "literally" to interpret Plato's demiurgical narrative is an ancient one, and the issue continues to be debated. For a relatively recent and to my mind persuasive account, see Aryeh Finkelberg, "Plato's Method in *Timaeus*," *American Journal of Philology* 117 (1996): 391–409. Finkelberg argues that Plato intended his linear chronological framing, together with the figure of the demiurge, to be understood as "methodological" devices or heuristic fictions.

13. Cf. the reference to "likely discourse" (ἐικός λόγος) in *Timaeus* 30b and 48d.

14. See Jacques Derrida, "The Animal That Therefore I Am (More to Follow)," in *The Animal That Therefore I Am*, ed. Marie-Louise Mallet, trans. David Wills (New York: Fordham University Press, 2008), 132.

15. Note that ancient commentators were perplexed that Timaeus refers to the divine demiurge as both "maker" and "father" (28c), given that these terms imply different forms of production. Contemporary commentators continue to puzzle over this as well. Matthias Vorwerk, for example, proposes that "in the *Timaeus* the Demiurge is Father in relation to the cosmos as a whole, namely as a living being, but Maker in relation to its constituent parts" ("Maker or Father? The Demiurge from Plutarch to Plotinus," in *One Book, the Whole Universe: Plato's Timaeus Today*, ed. Richard D. Mohr and Barbara M. Sattler [Las Vegas: Parmenides, 2010], 100).

16. Sallis, *Platonic Legacies*, 126.

17. See Sallis, *Chorology*, 84–85.

18. Sallis, *Platonic Legacies*, 127.

19. See Timothy Morton, *Hyperobjects: Philosophy and Ecology After the End of the World* (Minneapolis: University of Minnesota Press, 2013), 55–68.

20. See the careful discussion of this complex and contested passage in Sallis, *Chorology*, 101–6.

21. Sallis, *Chorology*, 105.

22. Sallis, *Chorology*, 114–15.

23. Sallis, *Chorology*, 118–19.

24. Sallis, *Chorology*, 118.

25. Morton, *Hyperobjects*, 11–12.

26. Graham Harman, "The Well-Wrought Broken Hammer: Object-Oriented Literary Criticism," *New Literary History* 43, no. 2 (2012): 188.

27. Responding to Harman's and Morton's championing of object-oriented ontologies at the expense of relational ontologies, Jane Bennett writes, "But perhaps there is no need to choose between objects or their relations" ("Systems and Things: A Response to Graham Harman and Timothy Morton," *New Literary History* 43 [2012]: 227). I am suggesting that thinking form via *khora* gives us both object and relation or flow.

28. Sallis, *Chorology*, 138.

29. Sallis, *Chorology*, 41.

30. Sallis, *Chorology*, 35.

31. Sallis, *Chorology*, 41.

32. Sallis, *Chorology*, 143.

33. Bennett, *Vibrant Matter*, 122.

34. John Dillon, "The *Timaeus* in the Old Academy," in *Plato's* Timaeus *as Cultural Icon*, ed. Gretchen J. Reydams-Schils (Notre Dame, IN: University of Notre Dame Press, 2003), 80.

35. Maren R. Niehoff, "Did the *Timaeus* Create a Textual Community?" *Greek, Roman, and Byzantine Studies* 47 (2007): 164–70.

36. Heinrich Dörrie, *Von Platon zum Platonismus: Ein Bruch in der Überlieferung und seine Überwindung*, Rheinisch-Westfälische Akademie der Wissenschaft Vorträge (Opladen, Germany: Westdeutscher Verlag, 1976), 32.

37. Carl Séan O'Brien, *The Demiurge in Ancient Thought: Secondary Gods and Divine Mediators* (Cambridge: Cambridge University Press, 2015), 307.

38. See Niehoff, "Did the *Timaeus* Create a Textual Community?" 162.

39. Sallis, *Platonic Legacies*, 4.

40. Niehoff, "Did the *Timaeus* Create a Textual Community?" 176, 174, 177.

41. David T. Runia, *Philo of Alexandria and the* Timaeus *of Plato* (Leiden, Netherlands: Brill, 1986), 523.

42. In this he was apparently anticipated by another Jewish-Alexandrian exegete, Aristobulus; see Eusebius, *Preparation for the Gospel* 13.12.

43. For Philo's *On the Creation*, I have used the Greek edition in Philo, *On the Creation: Allegorical Interpretation of Genesis 2 and 3*, trans. F. H. Colson and G. H. Whitaker, Loeb Classical Library 226 (Cambridge, MA: Harvard University Press, 1929); translations are my own. Note that Philo is only the second to use the term *cosmopolitan* in extant Greek literature, anticipating later Stoic usage. Before Philo, the Cynic Diogenes is said to have referred to himself as a "world-citizen," but his meaning seems to have been negative, i.e., a disavowal of his citizenship in any particular city (Diogenes Laertes 6.63). See David T. Runia, *Philo of Alexandria, On the Creation of the Cosmos According to Moses: Introduction, Translation, and Commentary* (Leiden, Netherlands: Brill, 2001), 103.

44. Runia notes that the phrase "maker and father" occurs twenty-one times in Philo's works and appears in the reverse order "father and maker" another twenty times, including *On the Creation* 10 and 12 (*Philo and the* Timaeus, 108).

45. Runia, *Philo and the* Timaeus, 109; Runia, *On the Creation*, 144–45.

46. Runia notes that Philo typically uses *ousia* as the functional equivalent of matter or stuff (Runia, *On the Creation*, 145).

47. Runia, *On the Creation*, 145.

48. Runia, *On the Creation*, 147.

49. As Runia notes, the question of whether Philo subscribes to the doctrine of *creatio ex nihilo* "is one of the most controversial topics in Philonic studies, on which so far no consensus has emerged" (Runia, *On the Creation*, 152).

50. Runia, *On the Creation*, 145.

51. Morton, *Ecological Thought*, 7.

52. In his dialogue *On Animals*, Philo repeats the Stoic argument in favor of anthropocentrism: humans are unique in their possession of reason; what appears as rationality in other animals is in fact mere instinct. He makes this case in response to his nephew Tiberius Julius Alexander's lengthy defense of the reasoned behavior of spiders, bees, swallows, monkeys, fawns, elephants, fish, tortoises, falcons, and others (see Abraham Terian, *Philonis Alexandrini de animalibus: The Armenian Text with an Introduction, Translation, and Commentary* [Chico, CA: Scholars Press, 1981], 67–108). David Clough notes that Philo's anthropocentric reading of Genesis based on the uniqueness of human rationality strongly influenced later Christian readers and that "the position he outlines broadly characterizes Christian readings of Genesis until the eighteenth century" ("All God's Creatures: Reading Genesis on Human and Non-Human Animals," in *Reading Genesis After Darwin*, ed. Stephen C. Barton and David Wilkinson [Oxford: Oxford University Press, 2009], 146). Clough himself argues against such "human separatism" based in part on a Darwinian perspective.

53. Although Origen names Philo only three times in his surviving corpus, it is clear that Philo's influence on Origen is pervasive. See Annawies van den Hoek, "Philo and Origen: A Descriptive Catalogue of Their Relationship," *Studia Philonica Annual* (2000): 44–121; van den Hoek, "Assessing Philo's Influence in Christian Alexandria: The Case of Origen," in *Shem in the Tents of Japhet: Essays on the Encounter of Judaism and Hellenism*, ed. James L. Kugel (Leiden, Netherlands: Brill, 2002), 223–39; and David T. Runia, "Philo and the Early Christian Fathers," in *The Cambridge Companion to Philo*, ed. Adam Kamesar (Cambridge: Cambridge University Press, 2009), 215–17.

54. For Origen's *Against Celsus*, I have used the edition in Origen, *Contre Celse*, ed. M. Borret, Sources Chrétiennes 132, 136, 147, 150, 227 (Paris: Cerf, 1967–76); translations generally follow Origen, *Contra Celsum*, trans. Henry Chadwick (Cambridge: Cambridge University Press, 1965).

55. See Porphyry, as cited by Eusebius, *Church History* 6.19.6–8.

56. Elizabeth DePalma Digeser, "Origen on the *Limes*: Rhetoric and the Polarization of Identity in the Late Third Century," in *The Rhetoric of Power in Late Antiquity: Religion and Politics in Byzantium, Europe and the Early Islamic World*, ed. Robert M. Frakes, Elizabeth DePalma Digeser, and Justin Stephens (London: Tauris Academic Studies, 2010), 204–5. See also Digeser, *A Threat to Public Piety: Christians, Platonists, and the Great Persecution* (Ithaca, NY: Cornell University Press, 2012), 49–71. Digeser has made a powerful case for the identity of what some have posited as two "Origens"—one the Platonist who studied with Ammonius and the other the Christian exegete. For a dissenting view, see Mark J. Edwards, "Ammonius, Teacher of Origen," *Journal of Ecclesiastical History* 44 (1993): 1–13.

57. Presumably Origen wants to show that Genesis attributes the creation of humans and even the lowliest animals to the god, in refuting Celsus's position. Since it is not clear what he gains in this context by pointing out that other parts of the cosmos are *not* said to have been created directly by the god, his point seems to be primarily methodological: an accurate reading of the biblical text must take account of its complexity. Note too that in his first Genesis Homily Origen parses the biblical text a bit differently, including only the heaven and earth, sun and moon, and humanity among those created directly by the god rather than merely at the god's command, with the stated aim of emphasizing the greatness of the human (Genesis Homily 1.12). The discrepancy is the result of the fact that in the case of the firmament and the animals of water, air, and earth, the biblical text includes both a command and the statement "God made."

58. See Cinzia Arruzza, "Plato's World-Maker in Origen's *Contra Celsum*," *Horizons* 3, no. 1 (2012): 71–72.

59. Enrica Ruaro, "'Resurrection: The Hope of Worms': The Dispute Between Celsus and Origen on the Resurrection of the Body," in *Perspectives sur le néoplatonisme: International Society of Neoplatonic Studies: Actes du colloque de 2006*, ed. Martin Achard, Wayne Hankey, and Jean-Marc Narbonne (Québec: Presses de l'Université Laval, 2009), 119–20.

60. Levi R. Bryant, *The Democracy of Objects* (Ann Arbor, MI: Open Humanities Press, 2011). As Catherine Keller reminds me and as Steven Shaviro also notes, long before Bryant, Alfred North Whitehead referred to "a democracy of fellow creatures" (*Process and Reality* [New York: Free Press, 1978; orig. 1929], 50; see Steven Shaviro, *The Universe of Things: On Speculative Realism* [Minneapolis: University of Minnesota Press, 2014], 60).

61. Arruzza, "Plato's World-Maker," 70.

62. Rufinus, for example, debates whether to translate the title as *De principiis* or *De principatibus* (Preface to On First Principles, 3). Origen himself discusses the broad semantic range of *arche* in *Commentary on John* 1.16.90–1.20.124, noting that "the Greeks are not alone in giving the term *arche* many meanings" (1.16.90).

63. Origen also refers to *archai* as doctrinal principles in his (recently discovered) *Homily 7 on Psalm 77*: "For there are many *archai* of doctrines: the *arche* about the father, the *arche* about the son, the *arche* about the holy spirit, the one about the church, the one about the holy powers" (Lorenzo Perrone, *Origenes XIII: Die neuen Psalmenhomilien: Eine kritische Edition des Codex Monacensis Graecus 314* [Berlin: Walter de Gruyter, 2015], 437). I thank Professor Perrone for pointing me to this passage.

64. For Origen's *On First Principles*, I have used the edition in Origen, *Traité des principes*, ed. Henry Crouzel and Manlio Simonetti, Sources Chrétiennes 252, 253, 268, 269, 312 (Paris: Cerf, 1978–84); translations generally follow Origen, *On First Principles*, trans. G. W. Butterworth (New York: Harper and Row, 1966).

65. See also Origen's *Genesis Homily* 1.1, where he opens by asking, "What is the beginning of all things except our lord and 'savior of all' (1 Tim. 4:10), Jesus Christ, 'the firstborn of every creature' (Col. 1:15)? In the beginning, therefore, that is, in his *logos*, 'God made heaven and earth' (Gen. 1:1)."

66. Patricia Cox Miller, *The Poetry of Thought in Late Antiquity: Essays in Imagination and Religion* (Aldershot, England: Ashgate, 2001), 183.

67. It must be said that the scholarly consensus is that neither the Greek fragment (which is found in Justinian's polemical *Letter to Mennas*) nor the other ancient witnesses to Origen's teaching of the transmigration of human souls to animals can be trusted, espe-

cially given Origen's explicit denials of the doctrine in his later works. On the latter, see Mario Maritano, "L'argomentazione scritturistica di Origene contro i sostenitori della metensomatosi," in *Origeniana sexta: Origen and the Bible*, ed. Gilles Dorival and Alain Le Boulluec (Leuven, Belgium: Peeters, 1995), 251–76, and Maritano, "Argomenti 'filosofici' di Origene contro la metensomatosi," in *Origeniana octava: Origen and the Alexandrian Tradition*, ed. Lorenzo Perrone (Leuven, Belgium: Peeters, 2002–3), 503–35. Nonetheless, the doctrine of which Origen was accused of holding is one that his thought might indeed seem to invite.

68. See Henry Chadwick, "Origen, Celsus and the Stoa," *Journal of Theological Studies* 48 (1947): 35–39. For a recent and concise account of the philosophical debate over animal rationality, see Janet E. Spittler, *Animals in the Apocryphal Acts of Apostles* (Tübingen, Germany: Mohr Siebeck, 2008), 15–26.

69. Elsewhere, Origen asserts that "all things are ensouled (πάντα ἐψύχωται) and there is nothing in the cosmos without soul; but all things are ensouled with different bodies. The heaven is ensouled, on account of which scripture speaks to it as to a living being (ὡς πρὸς ζῷον). . . . The earth is ensouled. . . . And so the heaven is ensouled and the earth is ensouled: are the seas and the rivers not ensouled, and indeed are not all things ensouled?" (Homily on Ps. 76, 3.2; Perrone, *Origenes XIII*, 329–30). I thank Professor Perrone for pointing me to this passage.

70. Carl O'Brien makes a similar, though not identical, point, suggesting that in Origen's philosophy "free will . . . takes the place of the recalicitrance of matter as the root of all evil" (O'Brien, *Demiurge in Ancient Thought*, 283). O'Brien's reading of both Plato and Origen is more sharply dualistic than my own, nor would I conflate *khora* with recalcitrant matter in the case of Plato.

71. Or is it Rufinus's language? Lorenzo Perrone points out to me that Rufinus often produces doublets not present in the original Greek. Obviously, I am pressing the text hard at this point in an attempt to draw out how Origen's cosmology holds unity and conflict together while also accommodating vast temporal and spatial scales.

72. Romans 8:22. We recall that Philo too refers to the earth as "travailing," albeit in a rather different context; Philo's earth gives birth to "every sort of cultivated plant, and every sort of tree, and also fruit in unspeakable abundance" (On the Creation 43).

73. See Calcidius, *Commentary on Plato's* Timaeus, 276–78; this section of the fourth-century Latin commentary appears to derive from Origen's lost *Commentary on Genesis*, which was written around the same time as *On First Principles* (Testimony C II 1 in Karin Metzler, ed. and trans., *Origenes: Die Kommentierung des Buches Genesis* [Berlin: De Gruyter, 2010], 46–52). See the discussion of J. C. M. van Winden, *Calcidius on Matter: His Doctrine and Sources—A Chapter in the History of Platonism* (Leiden, Netherlands: Brill, 1965), 52–66; and, more recently, Charlotte Köckert, *Christliche Kosmologie und kaiserzeitliche Philosophie: Die Auslegung des Schöpfungsberichtes bei Origenes, Basilius und Gregor von Nyssa vor dem Hintergrund kaiserzeitlicher Timaeus-Interpretationen* (Tübingen, Germany: Mohr Siebeck, 2009), 229–56. The unusual translation "speechless with amazement" seems to derive from psychologizing interpretations of the Hebrew *tohu va bohu* as "bewildered and astonished," indicating earth's response to find itself so disadvantaged in comparison with heaven (thus, Genesis Rabbah 2); Origen has shifted the interpretation of earth's response as one of stupefied admiration for its creator (J. C. M. van Winden, "'Terra autem stupida quadam erat admiratione': Reflections on a Remarkable Translation of

Genesis 1:2a," in *Arche: A Collection of Patristic Studies by J. C. M. van Winden*, ed. J. den Boeft and D. T. Runia [Leiden, Netherlands: Brill, 1977], 109–11).

74. See, for example, his allegorical interpretation of Genesis 1 in his first Genesis Homily.

75. H. Paul Santmire, *The Travail of Nature: The Ambiguous Ecological Promise of Christian Theology* (Philadelphia: Fortress, 1985), 52.

76. Timothy D. Barnes, *Athanasius and Constantius: Theology and Politics in the Constantinian Empire* (Cambridge, MA: Harvard University Press, 1993), 12. Despite his general conviction that Athanasius's level of education has been widely overestimated, Barnes here affirms the argument of E. P. Meijering, *Orthodoxy and Platonism in Athanasius: Synthesis or Antithesis?* (Leiden, Netherlands: Brill, 1968).

77. Rebecca Lyman makes the case that Athanasius's "work is no less cosmological or more soteriological than that of earlier authors," e.g., Origen (*Christology and Cosmology: Models of Divine Activity in Origen, Eusebius, and Athanasius*, Oxford Theological Monographs [Oxford: Clarendon, 1993], 125). Khaled Anatolios similarly emphasizes the "inner logic of the Athanasian vision" and foregrounds the significance of cosmology and the doctrine of creation (*Athanasius: The Coherence of His Thought* [London: Routledge, 1998], 1).

78. References are to the following edition and translation (though the translation is in some cases modified): Robert W. Thomson, ed. and trans., *Athanasius:* Contra gentes *and* De incarnatione (Oxford: Clarendon, 1971).

79. Cf. *Against the Gentiles* 34.3–4: "They receive the punishment for their folly by being numbered among the irrational animals."

80. The phrase is from Sallis, *Chorology*, 138. Plato refers to "creatures footless and crawling on the earth" (Timaeus 92a). Origen mentions snails among those seemingly insignificant creatures that nonetheless have souls (On First Principles 2.8.1).

81. Note that Athanasius's knowledge of the passage may be mediated by Eusebius; see Meijering, *Orthodoxy and Platonism*, 35–37.

82. Lyman, *Christology and Cosmology*, 138.

83. See Tertullian, *Against Hermogenes* 23.

84. As Lyman puts it, for Athanasius, "the innate weakness and instability of the creation . . . overshadow salvation history" and "will was not an important faculty, but was rather a largely negative expression of changeable nature" (Lyman, *Christology and Cosmology*, 139, 144).

85. Morton, *Ecological Thought*, 1.

86. David Brakke, "Athanasius," in *Augustine Through the Ages: An Encyclopedia* (Grand Rapids, MI: Eerdmans, 1999), 77.

87. György Heidl, *Origen's Influence on the Young Augustine: A Chapter in the History of Origenism* (Piscataway, NJ: Gorgias, 2003).

88. Heidl speculates that Simplicianus may have given Augustine a collection of Origen's writings (*Origen's Influence*, 35).

89. Cf. his earlier reference to *Plotini paucissimi libri* or "a few books of Plotinus" in *On the Blessed Life* 1.4 and his reference to the kindling of a ceaseless longing for "philosophy" and the "life that pleased and suited us" in *Against the Academics* 2.2.5.

90. Heidl, *Origen's Influence*, 35–36.

91. Catherine Keller, *The Face of the Deep: A Theology of Becoming* (London: Routledge, 2003), 75.

92. For Augustine's Genesis commentaries, I have used the edition in Augustine, *Opera omnia, tomus tertius*, Patrologia Latina 34 (1841); translations follow Augustine, *Saint Augustine: On Genesis*, trans. and ed. Edmund Hill (Hyde Park, NY: New City Press, 2002).

93. Augustine likely assumes, following Origen, that the term *heaven* refers to the invisible creation, the term *earth* to the visible one (Heidl, *Origen's Influence*, 90–91).

94. Cf. *Literal Meaning of Genesis* 3.3.4 on the four elements and debates about whether they change into one another or are stable.

95. Cf. *On Genesis Against the Manichaeans* 1.5.9 and *Unfinished Literal Commentary on Genesis* 3.10.

96. For Augustine's *Confessions*, I have used the edition in Augustine, *Confessions*, trans. Carolyn J.-B. Hammond, Loeb Classical Library 26 and 27 (Cambridge, MA: Harvard University Press, 2014–16); translations are my own.

97. Keller, *Face of the Deep*, 75.

98. Keller, *Face of the Deep*, 80.

99. Morton, *Ecological Thought*, 81. Of course, I am well aware that others have read Augustine differently, and not without reason. Andrea Nightingale, for example, asserts that "his discussions of mutability exhibit a sort of rage against time and the mortal body" (*Once Out of Nature: Augustine on Time and the Body* [Chicago: University of Chicago Press, 2011], 12).

100. H. L. Strack and Günter Stemberger, *Introduction to the Talmud and Midrash*, 2nd ed., trans. and ed. Markus Bockmuehl (Minneapolis: Fortress, 1996), 279–80.

101. Annette Yoshiko Reed, "Reading Augustine and/as Midrash: Genesis 6 in Genesis Rabbah and the City of God," in *Midrash and Context*, ed. Lieve M. Teugels and Rivka Ulmer (Piscataway, NJ: Gorgias, 2007), 107.

102. I follow the translation of Jacob Neusner, Genesis Rabbah, *the Judaic Commentary to the Book of Genesis: A New American Translation, Volume 1* (Atlanta: Scholars Press, 1985).

103. Peter Schäfer, "Bereshit Bara Elohim: Bereshit Rabba, Parashah 1, Reconsidered," in *Empsychoi Logoi—Religious Innovations in Antiquity: Studies in Honour of Pieter Willem van der Horst*, ed. Alberdina Houtman, Albert de Jong, and Magdalena Wilhelmina Misset-van de Weg (Leiden, Netherlands: Brill, 2008), 269.

104. Henry A. Fischel, "The Transformation of Wisdom in the World of Midrash," in *Aspects of Wisdom in Judaism and Early Christianity*, ed. Robert L. Wilken (Notre Dame, IN: University of Notre Dame Press, 1975), 80.

105. Maren Niehoff, "*Creatio ex Nihilo* Theology in Genesis Rabbah in Light of Christian Exegesis," *Harvard Theological Review* 99, no. 1 (2005): 61–64.

106. The Mishnah and Tosefta restrict the study of three topics: illicit sexual relations as discussed in Leviticus 18 and 20; the works of creation as described in Genesis 1; and speculations about the godhead and divine realm based on Ezekiel 1 (Mishnah Hagigah 2:1, Tosefta Hagigah 2:1–7). The origins and significance of this tradition have been much discussed and debated. Some recent treatments include Annette Yoshiko Reed, "From 'Pre-Emptive Exegesis' to 'Pre-Emptive Speculation'? *Ma'aseh Bereshit* in *Genesis Rabbah* and *Pirqei DeRabbi Eliezer*," in *With Letters of Light/Otiyot Shel or: Studies in Early Jewish Apocalypticism and Mysticism in Honour of Rachel Elior*, ed. D. Arbel and A. Orlov (Berlin: De Gruyter, 2010), 115–32; Yair Furstenberg, "The Rabbinic Ban on *Ma'aseh Bereshit*: Sources, Contexts," in *Jewish and Christian Cosmogony in Late Antiquity*, ed. Lance Jenott and Sarit

Kattan Gribetz (Tübingen, Germany: Mohr Siebeck, 2013), 39–63; Reimund Leicht, "Major Trends in Rabbinic Cosmology," in *Hekhalot Literature in Context: Between Byzantium and Babylonia*, ed. Ra'anon Boustan, Martha Himmelfarb, and Peter Schäfer (Tübingen, Germany: Mohr Siebeck), 245–78.

107. See the discussion of Gershom Scholem, *Jewish Gnosticism, Merkavah Mysticism, and Talmudic Tradition*, 2nd rev. ed. (New York: Jewish Theological Seminary, 1965), 36–42.

108. Cf. Mishnah Hagigah 2.1.

109. Niehoff, "*Creatio ex Nihilo* Theology," 46–47.

110. Niehoff, "*Creatio ex Nihilo* Theology," 64. See also the alternate account of Menahem Kister, who is inclined to see the Jewish rejection of a concept of uncreated matter as (1) emerging earlier and (2) being generated less by polemics, rivalry, or external influence than by "an intrinsic exegetical and theological problem arising out of Genesis 1:2" (Menahem Kister, "Tohu Wa-Bohu: Primordial Elements and Creatio Ex Nihilo," *Jewish Studies Quarterly* 14 [2007]: 256).

111. Schäfer, "Bereshit Bara Elohim," 287.

112. Leicht, "Major Trends in Rabbinic Cosmology," 259–60.

113. Sallis, *Chorology*, 58.

114. Richard Kearney, *The God Who May Be: A Hermeneutics of Religion* (Bloomington: Indiana University Press, 2001).

115. John D. Caputo, *The Insistence of God: A Theology of Perhaps* (Bloomington: Indiana University Press, 2013), 13.

116. John D. Caputo, *The Weakness of God: A Theology of the Event* (Bloomington: Indiana University Press, 2006), 64.

117. John D. Caputo, *The Prayers and Tears of Jacques Derrida: Religion Without Religion* (Bloomington: Indiana University Press, 1997), 36.

118. Caputo, *Insistence of God*, 190.

119. Keller, *Face of the Deep*, 219.

120. Keller, *Face of the Deep*, 232.

121. Keller, *Face of the Deep*, 231.

122. Mary-Jane Rubenstein, "Cosmic Singularities: On the Nothing and the Sovereign," *Journal of the American Academy of Religion* 80, no. 2 (2012): 488.

123. Keller, *Face of the Deep*, offers a thoroughgoing and distinctly feminist critique of *creatio ex nihilo*. Caputo echoes her critique, focusing on the problem of theodicy, declaring that the concept of divine omnipotence that emerged in antiquity "had a tin ear for life's contingencies and would thereafter have the effect of laying the horrors of this life squarely at the feet of God" (Caputo, *Weakness of God*, 75). For a range of theological responses to the doctrine, see Janet Martin Soskice, ed., "Creation 'ex Nihilo' and Modern Theology," *Modern Theology* 29, no. 2 (2013); and Thomas Jay Oord, ed., *Theologies of Creation:* Creatio ex Nihilo *and Its New Rivals* (New York: Routledge, 2015). Particularly relevant is Kathryn Tanner's contribution to the former volume. Tanner links creation ex nihilo with the incomplete rejection of both "creation from or through something" (the demiurgic model) and "creation out of god" (the emanationist or erotic model). "Typically creation ex nihilo positions include the imagery and concepts of both these rejected viewpoints . . . while violating their proper bounds." In her view, hierarchical or dominative thinking is not inherent to the ex nihilo: while it "may be linked to what Catherine Keller calls *tehomophobia*, fear of the deep with all its gendered resonances, . . . it need

not" (Tanner, "*Creatio Ex Nihilo* as Mixed Metaphor," *Modern Theology* 29, no. 2 [2013]: 149, 140).

124. Sallis, *Chorology*, 123.

125. Keller, *Face of the Deep*, xvi.

126. Chris Boesel and Catherine Keller, "Introduction," in *Apophatic Bodies: Negative Theology, Incarnation, and Relationality*, ed. Chris Boesel and Catherine Keller (New York: Fordham University Press, 2010), 11–12.

127. Diana Coole and Samantha Frost, "Introducing the New Materialisms," in *New Materialisms: Ontology, Agency, and Politics*, ed. Diana Coole and Samantha Frost (Durham, NC: Duke University Press, 2010), 7.

128. Bryant, *Democracy of Objects*.

129. Morton, *Ecological Thought*, 132–33.

130. Morton, *Ecological Thought*, 124.

131. Catherine Keller, *Cloud of the Impossible: Negative Theology and Planetary Entanglement* (New York: Columbia University Press, 2015), 269.

132. Keller, *Cloud of the Impossible*, 307.

133. Keller, *Cloud of the Impossible*, 289.

134. Keller, *Cloud of the Impossible*, 286.

135. Keller, *Cloud of the Impossible*, 316.

136. Sharon V. Betcher, *Spirit and the Obligation of Social Flesh: A Secular Theology for the Global City* (New York: Fordham University Press, 2014), 12. See also Sharon V. Betcher, "Crip/Tography: Disability Theology in the Ruins of God," *Journal for Cultural and Religious Theory* 14, no. 3 (2015): 114.

137. Betcher, *Spirit and the Obligation of Social Flesh*, 190–91.

138. Betcher, *Spirit and the Obligation of Social Flesh*, 17.

139. Betcher, "Crip/Tography," 112.

140. Betcher, *Spirit and the Obligation of Social Flesh*, 17–18.

141. Keller, *Cloud of the Impossible*, 10.

142. Keller, *Cloud of the Impossible*, 302.

143. Claire Colebrook, "We Have Always Been Post-Anthropocene: The Anthropocene Counter-Factual," academia.edu (2014), 4, https://www.academia.edu/12757260/We_Have_Always_Been_Post-Anthropocene.

144. Morton, *Ecological Thought*, 1.

II. Queering Creation: Hagiography Without Humans

Note to epigraph: Morton, "Queer Ecology," 277–78, 280.

1. Morton, "Queer Ecology," 273. See also Morton, *Ecological Thought*, 81–87.

2. See especially Bruce Erickson and Catriona Mortimer-Sandilands, "Introduction: A Genealogy of Queer Ecologies," in *Queer Ecologies: Sex, Nature, Politics, Desire*, ed. Bruce Erickson and Catriona Mortimer-Sandilands (Bloomington: Indiana University Press, 2010), 1–47. See also Catriona Mortimer-Sandilands, "Unnatural Passions? Notes Toward a Queer Ecology," *Invisible Culture: An Electronic Journal for Visual Culture* 9 (2005), https://www.rochester.edu/in_visible_culture/Issue_9/sandilands.html.

3. Alex Johnson, "How to Queer Ecology: One Goose at a Time," *Orion Magazine* 30, no. 2 (2011): 46–49.

4. Greg Garrard, "How Queer Is Green?" *Configurations* 18, no. 1/2 (2010): 73–96.

5. The special issue of the journal *Worldviews: Global Religions, Culture, and Ecology*, dedicated to the topic of "Religion, Disability, and the Environment," showcases the rich intersections of ecological and disability theories; queer theory is also invoked by both Whitney Bauman ("Disability Studies, Queer Theory, and the New Materialism," *Worldviews* 19, no. 1 [2015]: 69–73) and Mel Chen ("The Reproduction in/of Disability and Environment," *Worldviews* 19, no. 1 [2015]: 79–82). See also Mel Y. Chen, *Animacies: Biopolitics, Racial Mattering, and Queer Affect* (Durham, NC: Duke University Press, 2012), a work that locates itself at the intersection of new materialisms, queer theory, critical race theory, critical animal studies, and disability theory; and Whitney A. Bauman, *Religion and Ecology: Developing a Planetary Ethic* (New York: Columbia University Press, 2014), in which queer theory plays a significant role in destabilizing notions of nature and identity and challenging human exceptionalism.

6. See Catriona Sandilands, "Lavender's Green? Some Thoughts on Queer(y)ing Environmental Politics," *Undercurrents* 6 (1994): 20–25.

7. Greta Gaard, "Toward a Queer Ecofeminism," *Hypatia* 12, no. 1 (1997): 114–37.

8. Morton, *Ecological Thought*, 5.

9. Morton, *Ecological Thought*, 5.

10. See Sharon V. Betcher, "The Picture of Health: 'Nature' at the Intersection of Disability, Religion, and Ecology," *Worldviews* 19, no. 1 (2015): 9–33.

11. Donna J. Haraway, *The Companion Species Manifesto: Dogs, People, and Significant Others* (Chicago: University of Chicago Press, 2003).

12. Timothy Morton, *Ecology Without Nature: Rethinking Environmental Aesthetics* (Cambridge, MA: Harvard University Press, 2009).

13. Morton, "Queer Ecology," 275–76.

14. See Bruce Bagemihl, *Biological Exuberance: Animal Homosexuality and Natural Diversity* (New York: St. Martin's, 1999); Joan Roughgarden, *Evolution's Rainbow: Diversity, Gender, and Sexuality in Nature and People* (Berkeley: University of California Press, 2004); Paul Vasey and Volker Sommer, eds., *Homosexual Behaviour in Animals: An Evolutionary Perspective* (Cambridge: Cambridge University Press, 2006).

15. Morton, *Ecological Thought*, 61.

16. Erickson and Mortimer-Sandilands, "Genealogy of Queer Ecologies," 13.

17. Morton, *Ecological Thought*, 81.

18. Mayra Rivera, *Poetics of the Flesh* (Durham, NC: Duke University Press, 2015).

19. Virginia Burrus, *The Sex Lives of Saints: An Erotics of Ancient Hagiography*, Divinations: Rereading Late Ancient Religion (Philadelphia: University of Pennsylvania Press, 2004).

20. The language of "strange stranger" is Morton's translation of Derrida's *arrivant* (*Ecological Thought*, 140, n. 39). Cf. Eric Santner's concept of the "creature," a development of Franz Rosenzweig's concept of the "neighbor": "The new ethics of neighbor love adumbrated by Rosenzweig locates our responsibility in our capacity to elaborate forms of solidarity with this creaturely expressivity that makes the other strange not only to me but also to him- or herself" (Santner, *On Creaturely Life: Rilke, Benjamin, Sebald* [Chicago: University of Chicago Press, 2006], xiii).

21. Edward O. Wilson, "Biophilia and the Conservation Ethic," in *The Biophilia Hypothesis*, ed. Stephen R. Kellert and Edward O. Wilson (Washington, DC: Island, 1993), 31. "Nor is biophilia limited to humans," as Donovan Schaefer notes (*Religious Affects*, 197).

On naturalist Ellen Meloy's articulation of a distinctly queer biophilia, see Dianne Chisholm, "Biophilia, Creative Involution, and the Ecological Future of Queer Desire," in *Queer Ecologies: Sex, Nature, Politics, Desire*, ed. Catriona Mortimer-Sandilands and Bruce Erickson (Bloomington: Indiana University Press, 2010), 359–81.

22. Tobin Siebers, *Disability Aesthetics* (Ann Arbor: University of Michigan Press, 2010).

23. Timothy Morton, "From Modernity to the Anthropocene: Ecology and Art in the Age of Asymmetry," *International Social Science Journal* 63, no. 207/208 (2012): 46, 49.

24. Jacques Derrida, *The Animal That Therefore I Am*, ed. Marie-Louise Mallet, trans. David Wills (New York: Fordham University Press, 2008), 132.

25. Morton, "Queer Ecology," 278.

26. Derrida, *Animal That Therefore I Am*, 5–6.

27. Derrida, *Animal That Therefore I Am*, 119.

28. Derrida, *Animal That Therefore I Am*, 3.

29. Derrida, *Animal That Therefore I Am*, 10.

30. Derrida, *Animal That Therefore I Am*, 65.

31. Derrida, *Animal That Therefore I Am*, 10.

32. Derrida, *Animal That Therefore I Am*, 85.

33. Derrida, *Animal That Therefore I Am*, 132, 41.

34. Patricia Cox (Miller), *Biography in Late Antiquity: A Quest for the Holy Man*, Transformation of the Classical Heritage (Berkeley: University of California Press, 1983), 110, 107.

35. Derek Krueger, *Writing and Holiness: The Practice of Authorship in the Early Christian East*, Divinations: Rereading Late Ancient Religion (Philadelphia: University of Pennsylvania Press, 2004), 135.

36. For Porphyry's *Life of Plotinus*, I have used the edition in Plotinus, *Volume I: Porphyry on the Life of Plotinus. Ennead I*, trans. A. H. Armstrong, Loeb Classical Library 440 (Cambridge, MA: Harvard University Press, 1969); translations are my own.

37. Cox (Miller), *Biography in Late Antiquity*, 103.

38. Pierre Hadot, *Philosophy as a Way of Life: Spiritual Exercises from Socrates to Foucault*, ed. Arnold Davidson, trans. Michael Chase (Oxford: Blackwell, 1995; French, 1981), 201. The statement is applied to Plotinus by Margaret R. Miles, *Plotinus on Body and Beauty* (Oxford: Blackwell, 1999), xi.

39. The clearest reference is to Porphyry's *Life of Pythagoras*, but it seems to me not unlikely that Athanasius (assuming he is indeed the author of the *Life of Antony*, a debated issue) has his *Life of Plotinus* in mind as well; see Virginia Burrus, "Bodies, Desires, Confessions: Shame in Plotinus, Antony, and Augustine," in *Shame Between Penance and Punishment*, ed. Bénédicte Sère and Jörg Wettlaufer (Florence: SISMEL-Edizioni del Galluzzo, 2013), 34, n. 19.

40. The *Life of Mary of Egypt* is traditionally attributed to Sophronius, who was bishop of Jerusalem 634 to 638; not all scholars accept this attribution, but authorship is not significant for my argument. See Maria Kouli, "Life of St. Mary of Egypt," in *Holy Women of Byzantium: Ten Saints' Lives in English Translation*, ed. Alice-Mary Talbot (Washington, DC: Dumbarton Oaks Research Library and Collection, 1996), 66.

41. Elizabeth Freeman, "Time Binds, or, Erotohistoriography," *Social Text* 23, no. 3–4(84–85) (2005): 66.

42. Judith Halberstam, *In a Queer Time and Place: Transgender Bodies, Subcultural Lives* (New York: New York University Press, 2005), 2.

43. Halberstam, *In a Queer Time and Place*, 153.

44. Halberstam, *In a Queer Time and Place*, 2.

45. For the *Life of Antony*, I have used the edition in Athanasius, *Vie d'Antoine*, ed. G. J. M. Bartelink, Sources Chrétiennes 400 (Paris: Cerf, 1994); translations generally follow Athanasius, *The Life of Antony and the Letter to Marcellinus*, trans. and intro. Robert C. Gregg, Classics of Western Spirituality (New York: Paulist, 1980).

46. Halberstam, *In a Queer Time and Place*, 13.

47. Edith Wyschogrod, *Saints and Postmodernism: Revisioning Moral Philosophy* (Chicago: University of Chicago Press, 1990), 83. I discuss Wyschogrod's position at more length in Virginia Burrus, "Wyschogrod's Hand: Saints, Animality, and the Labor of Love," *Philosophy Today* 55, no. 4 (2011): 412–21.

48. For Jerome's *Life of Paul*, I have used the edition in W. Oldfather et al., *Studies in the Text Traditions of St. Jerome's* Vitae Patrum (Urbana: University of Illinois Press, 1943); translations generally follow Paul B. Harvey, "Jerome, 'Life of Paul, the First Hermit,'" in *Ascetic Behavior in Greco-Roman Antiquity: A Sourcebook*, ed. Vincent L. Wimbush (Minneapolis: Fortress, 1990), 357–69.

49. For the *Life of Mary of Egypt*, I have used the edition in Procopius of Gaza et al., *Opera omnia, tomus tertius*, Patrologia Graeca 87.3 (1865); translations follow Kouli, "Life of St. Mary of Egypt."

50. Morton, *Ecological Thought*, 98.

51. Morton, *Ecological Thought*, 119.

52. Shannon Winnubst, *Queering Freedom* (Bloomington: Indiana University Press, 2006), 190, 139.

53. Nicole Seymour, *Strange Natures: Futurity, Empathy, and the Queer Ecological Imagination* (Champaign: University of Illinois Press, 2013), 17.

54. Catriona Sandilands, "Queer Life? Ecocriticism After the Fire," in *The Oxford Handbook of Ecocriticism*, ed. Greg Garrard (Oxford: Oxford University Press, 2014), 310.

55. Elizabeth Freeman, *Time Binds: Queer Temporalities, Queer Histories* (Durham: Duke University Press, 2010), xvi.

56. Ernst Robert Curtius, *European Literature and the Latin Middle Ages*, trans. Willard R. Trask (Princeton, NJ: Princeton University Press, 1953; German original, 1948), 195.

57. Mark J. Edwards, "*Locus Horridus* and *Locus Amoenus*," in Homo Viator: *Classical Essays for John Bramble*, ed. Michael Whitby, Philip Hardie, and Mary Whitby (Bristol, UK: Bristol Classical Press, 1987), 267–76.

58. Curtius, *European Literature*, 195.

59. Alfred K. Siewers, *Strange Beauty: Ecocritical Approaches to Early Medieval Landscape* (New York: Palgrave Macmillan, 2009), 11.

60. Sara Ahmed, "Happy Objects," in *The Affect Theory Reader*, ed. Melissa Gregg and Gregory J. Seigworth (Durham, NC: Duke University Press, 2010), 29.

61. Ahmed, "Happy Objects," 30.

62. Lawrence Buell, *The Future of Environmental Criticism: Environmental Crisis and Literary Imagination* (Oxford: Blackwell, 2005), 72–73.

63. Scott Hess, "Postmodern Pastoral, Advertising, and the Masque of Technology," *Interdisciplinary Studies in Literature and Environment* 11 (2004): 71–100.

64. Morton, *Ecological Thought*, 50.

65. Morton, *Ecological Thought*, 52.

66. Petra Hass, *Der* locus amoenus *in der antiken Literatur: Zu Theorie und Geschichte eines literarischen Motivs* (Bamberg, Germany: Wissenschaftler Verlag, 1998), 98.

67. Morton, *Ecological Thought*, 83.

68. Morton, *Ecological Thought*, 83.

69. Aneta Kliszcz and Joanna Komorowska, "Glades of Dread: The Ecology and Aesthetics of *Loca Horrida*," in *Ecocriticism, Ecology, and the Cultures of Antiquity*, ed. Christopher Schliephake (Lanham, MD: Lexington, 2017), 53–54.

70. Curtius, *European Literature*, 196.

71. Long before Jerome, Ovid revised the depiction of the *locus amoenus* as a safe place; scenes of sexual violence repeatedly disrupt the peace of the pastoral setting: "The horror of the violent act itself is accentuated by the concomitant betrayal of the promise of security implied by the beauty and isolation of the *locus amoenus*" (Neil W. Bernstein, "*Locus Amoenus* and *Locus Horridus* in Ovid's *Metamorphoses*," *Wenshan Review of Literature and Culture* 5 [2011]: 83). For Jerome, however, this is crucially a scene of seduction rather than rape (however blurry the lines may become) because it must function as a temptation. The *locus amoenus* is not a safe space betrayed but is, rather, inherently unsafe because of its seductive pleasures.

72. Compare Hilarion's own dwelling in mountainous Cyprus, as Jerome depicts it in his *Life of Hilarion*, a "very pleasant little garden" (*hortulus peramoenus*) with fruit trees and a stream that abuts the ruins of a pagan temple—perhaps that of the goddess of love herself—and is rather literally haunted; it is a "terrible and remote place . . . where the voices of demons resounded night and day" that delights Hilarion nonetheless (Life of Hilarion 43). Here too there is a queer convergence of the *locus amoenus* and the *locus horridus*, in a happy place that has become an uncanny place that itself makes the hermit happy.

73. Note that long-standing traditions of literary pastoral have contributed to the association of the rural landscape not only with beauty and leisure but also with an uncultivated or "natural" heterosexual manhood. However, literary traditions of "gay pastoral" are also long-standing, harking back to models as ancient as Theocritus's *Idylls* and Virgil's *Eclogues*—which, as David Shuttleton points out, "portray a polymorphous range of desires, ranging from male-male love-elegy to heterosexual romance and bawdy bestiality" yet still may leave "other repressive technologies of privilege, privatization and exclusion" intact ("The Queer Politics of Gay Pastoral," in *De-Centering Sexualities: Politics and Representations Beyond the Metropolis*, ed. Richard Phillips, Diane Watt, and David Shuttleton [London: Routledge, 2000], 125, 140).

74. Greg Garrard, "Radical Pastoral?" *Studies in Romanticism* 35, no. 3 (1996): 464.

75. Lance Newman and Laura Walls, "Cosmopolitics and the Radical Pastoral: A Conversation with Lawrence Buell, Hsuan Hsu, Antony Lioi, and Paul Outka," *Journal of Ecocriticism* 3, no. 2 (2011): 59–60.

76. Newman and Walls, "Cosmopolitics and the Radical Pastoral," 61.

77. Newman and Walls, "Cosmopolitics and the Radical Pastoral," 62, 68.

78. Bauman, *Religion and Ecology*, 13.

79. Patricia Cox Miller, "Adam, Eve, and the Elephants: Asceticism and Animality," in *Ascetic Culture: Essays in Honor of Philip Rousseau*, ed. Blake Leyerle and Robin Darling Young (Notre Dame, IN: University of Notre Dame Press, 2013), 265.

80. David Brakke, *Demons and the Making of the Monk: Spiritual Combat in Early Christianity* (Cambridge, MA: Harvard University Press, 2006), 158, 181.

81. Jean-Christophe Bailly, *The Animal Side*, trans. Catherine Porter (New York: Fordham University Press, 2011), 12.

82. See the illumining and groundbreaking treatment of this centaur by Patricia Cox Miller, "Jerome's Centaur: A Hyper-Icon of the Desert," *Journal of Early Christian Studies* 4, no. 2 (1996): 209–33.

83. Felicitiously, the word *grex* can apply to a flock or herd, or by extension to human groups.

84. Patricia Cox Miller, "Is There a Harlot in This Text? Hagiography and the Grotesque," in *The Cultural Turn in Late Ancient Studies: Gender, Asceticism, and Historiography*, ed. Dale B. Martin and Patricia Cox Miller (Durham, NC: Duke University Press, 2005), 88–102.

85. Morton, "Queer Ecology," 277.

86. Tim Ingold, *The Perception of the Environment: Essays on Livelihood, Dwelling and Skill* (London: Routledge, 2000), 191.

87. Ingold, *Perception of the Environment*, 200.

88. Schaefer, *Religious Affects*, 162.

89. François Zourabichvili, "Six Notes on the Percept (on the Relation Between the Critical and the Clinical)," in *Deleuze: A Critical Reader*, ed. Paul Patton (Cambridge, MA: Blackwell, 1996), 196.

90. M. Tengberg, "Beginnings and Early History of Date Palm Garden Cultivation in the Middle East," *Journal of Arid Environments* 86 (2012): 139–47.

91. The Kitchen Sisters and Lisa Morehouse, "Forbidding Fruit: How America Got Turned on to the Date," National Public Radio, *Morning Edition*, June 10, 2014, http://www.npr.org/sections/thesalt/2014/06/10/320346869/forbidding-fruit-how-america-got-turned-on-to-the-date.

92. John Roach, "'Methuselah' Palm Grown from 2,000-Year-Old Seed is a Father," *National Geographic*, March 25, 2015, http://news.nationalgeographic.com/2015/03/150324-ancient-methuselah-date-palm-sprout-science/.

93. See Amanat Ali et al., "Nutritional and Medicinal Value of Date Fruit," in *Dates: Production, Processing, Food, and Medicinal Values*, ed. A. Manickavasagan, M. Mohamed Essa, and E. Sukumar (Boca Raton, FL: CRC Press, 2012), 361–75.

94. Morton, *Dark Ecology*, 42.

95. "Israeli Dates" (web page), BDSlist.com, accessed May 24, 2017, http://bdslist.org/consumer/israeli-dates/.

96. Dalia Hatuqa, "Palestinian Farmers Thirsty for Exports," *Al Jazeera*, April 10, 2013, accessed May 31, 2017, http://www.aljazeera.com/indepth/features/2013/04/20134312648559299.html.

97. See Susan Weingarten, *The Saint's Saints: Hagiography and Geography in Jerome* (Leiden, Netherlands: Brill, 2005), 58.

98. Theresa Gross-Diaz, "Cat.22: Tetradrachm (Coin) Portraying Queen Cleopatra VII, 37/33 B.C," in *Roman Art at the Art Institute of Chicago*, ed. Katharine A. Raff (Chicago: Art Institute of Chicago, 2017), accessed June 4, 2017, https://publications.artic.edu/roman/reader/romanart/section/510/510_anchor.

99. Weingarten, *Saint's Saints*, 30–31.

100. W. V. Harris, "Defining and Detecting Mediterranean Deforestation, 800 B.C.E to 700 C.E.," in *The Ancient Mediterranean Environment Between Science and History*, ed. W. V. Harris (Leiden, Netherlands: Brill, 2013), 176, 188.

101. Zekai Şen, *Wadi Hydrology* (Boca Raton, FL: CRC Press, 2008), xiii.

102. Şen, *Wadi Hydrology*, 178.

103. Şen, *Wadi Hydrology*, 98.

104. For the *Life of Syncletica*, I have used the edition in Athanasius, *Opera omnia, tomus quartus*, Patrologia Graeca 28 (1887); translations follow Elizabeth A. Castelli, "Life of Syncletica," in *Ascetic Behavior in Greco-Roman Antiquity*, ed. Vincent L Wimbush (Minneapolis: Fortress, 1990), 256–311.

105. Lynda L. Coon, *Sacred Fictions: Holy Women and Hagiography in Late Antiquity*, Middle Ages Series (Philadelphia: University of Pennsylvania Press, 1997), 40.

106. Betcher, *Spirit and the Obligation of Social Flesh*, 115.

107. Betcher, *Spirit and the Obligation of Social Flesh*, 108–9.

108. Rosemarie Garland-Thomson, "The Case for Conserving Disability," *Bioethical Inquiry* 9 (2012): 342.

109. Siebers, *Disability Aesthetics*, 10.

110. Siebers, *Disability Aesthetics*, 4.

111. Siebers, *Disability Aesthetics*, 5.

112. Betcher, *Spirit and the Obligation of Social Flesh*, 60.

113. Morton, *Ecological Thought*, 277.

114. For both Theodoret's and Antonius's *Lives of Simeon*, I use the editions in H. Lietzmann, *Das Leben des heiligen Symeon Stylites*, Texte und Untersuchung zur Geschichte der altchristlichen Literatur 32.4 (Leipzig: Hinrichs, 1908); translations follow Robert Doran, *The Lives of Simeon Stylites* (Kalamazoo, MI: Cistercian Publications, 1992).

115. Susan Ashbrook Harvey, "The Sense of a Stylite: Perspectives on Simeon the Elder," *Vigiliae Christianae* 42, no. 4 (1988): 378.

116. As Susan Harvey argues, with reference to this passage, "Simeon represents a mirrored image of his [Theodoret's] Christological position" ("Sense of a Stylite," 380).

117. Donna J. Haraway, *Simians, Cyborgs, and Women: The Reinvention of Nature* (New York: Routledge, 1991), 149.

118. Richard Grusin, "Introduction," in *The Nonhuman Turn*, ed. Richard Grusin (Minneapolis: University of Minnesota Press, 2015), viii–ix.

119. Cary Wolfe, "Introduction: What Is Posthumanism?" in *What Is Posthumanism?* ed. Cary Wolfe (Minneapolis: University of Minnesota Press, 2010), xv–xvi.

120. Bennett, *Vibrant Matter*, 31.

121. Becoming-other has its conceptual matrix in Deleuze and Guattari's concept of *becoming-animal*. As Matthew Calarco puts it, "Becoming-other is a refusal to enact the ideals and subjectivity that the dominant culture associates with being a full human subject and to enter into a relation with the various minor, or nondominant, modes of existence that are commonly viewed as being the 'other' of the human" (*Thinking Through Animals: Identity, Difference, Indistinction* [Stanford, CA: Stanford University Press, 2015], 57).

122. Glenn Peers, "Byzantine Things in the World," in *Byzantine Things in the World*, ed. Glenn Peers (New Haven, CT: Yale University Press, 2013), 74–75.

123. At this point I should acknowledge that others have argued that Theodoret at least is precisely interested in such an argument; see Harvey, "Sense of a Stylite," 379. If so, Simeon has escaped the control of the text.

124. On the icon as part of Simeon's "extended body," see Peers, "Byzantine Things," 73.

125. Gilles Deleuze and Felix Guattari, *A Thousand Plateaus: Capitalism and Schizophrenia*, trans. Brian Massumi (Minneapolis: University of Minnesota Press, 1987).

126. The column of course can also be interpreted in phallic terms, as can the ladder. The sinuousness of the serpent contrasts with the austere erectness of the column and ladder.

127. On hagiography and biblical typology more generally, see Derek Krueger, "Typological Figuration in Theodoret of Cyrrhus's Religious History and the Art of Postbiblical Narrative," *Journal of Early Christian Studies* 5, no. 3 (1997): 393–419.

128. On "the force of things," see Bennett, *Vibrant Matter*, 1–19. On the force of ancient Christian things in particular, see Glenn Peers, "Object Relations: Theorizing the Late Antique Viewer," in *The Oxford Handbook of Late Antiquity*, ed. Scott Fitzgerald Johnson (Oxford: Oxford University Press, 2012), 970–87; and Peers, "Byzantine Things."

129. Rope is rush (though Simeon's rope is made of palm leaves). Papyrus is a rushlike plant.

130. "Orlan: Performance" (web page), Orlan.eu, accessed June 13, 2017, http://www.orlan.eu/works/performance-2/. See also Tracey Warr, ed., *The Artist's Body* (London: Phaidon, 2000), 185.

131. Athey's work has received extensive treatment in Dominic Johnson, ed., *Pleading in the Blood: The Art and Performances of Ron Athey* (Bristol, UK: Intellect, 2013).

132. "Marina Abramovic: *Seven Easy Pieces*" (web page), Guggenheim Museum, accessed June 13, 2017, http://pastexhibitions.guggenheim.org/abramovic/.

133. "American Reflexxx," directed by Alli Coates, accessed June 13, 2017, https://www.youtube.com/watch?v=bXn1xavynj8.

134. Leslie Hill and Helen Paris, *The Guerilla Guide to Performance Art* (London: Continuum, 2001), 62.

135. Warr, *Artist's Body*, 101.

136. Carolyn Black, "Gina Pane: Arnolfini, Bristol, 23.02.02 to 13.04.02," *Decide Magazine* 2 (2002): 11.

137. Warr, *Artist's Body*, 130.

138. Warr, *Artist's Body*, 131.

139. See *Sick: The Life and Death of Bob Flanagan, Supermasochist*, documentary, dir. Kirby Dick, released November 7, 1997.

140. Amelia Jones, "Everybody Dies . . . Even the Gorgeous: Resurrecting the Work of Hannah Wilke," *Markszine* (2003): 20.

141. Jones, "Everybody Dies," 22.

142. Alphonso Lingis, "Animal Body, Inhuman Face," in *Zoontologies: The Question of the Animal*, ed. Cary Wolfe (Minneapolis: University of Minnesota Press, 2003), 168.

143. Bailly, *Animal Side*, 64.

144. Ellen Meloy, *The Anthropology of Turquoise: Reflections on Desert, Sea, Stone, and Sky* (New York: Random House, 2002), 221, cited by Chisholm, "Biophilia," 365.

145. Dorothy Allison, *Skin: Talking About Sex, Class, and Literature* (Ithaca, NY: Firebrand, 1994), 250.

146. Morton, *Ecological Thought*, 280.

147. Betcher, "Picture of Health," 11, 12, 28, 23.

III. Things and Practices: Arts of Coexistence

Note to epigraph: Bennett, *Vibrant Matter*, ix.

1. Bill Brown, "Thing Theory," *Critical Inquiry* 28, no. 1 (2001): 1–22.

2. Martin Heidegger, "The Thing," in *Poetry, Language, Thought*, trans. Albert Hofstadter (New York: Harper and Row, 1971), 165.

3. Heidegger, "The Thing," 172, 175.

4. Graham Harman, *Heidegger Explained: From Phenomenon to Thing* (Chicago: Open Court, 2007), 130, 161.

5. Heidegger, "The Thing," 180.

6. For critical engagement of Heidegger on "the animal," see Jacques Derrida, "I Don't Know Why We Are Doing This," in *Animal That Therefore I Am*, 141–60; and Graham Harman, *Tool-Being: Heidegger and the Metaphysics of Objects* (Chicago: Open Court, 2002), 69–80.

7. Shaviro, *Universe of Things*, 1.

8. Others include Ian Bogost, *Alien Phenomenology, or What It's Like to Be a Thing* (Minneapolis: University of Minnesota Press, 2012); Bryant, *Democracy of Objects*, and *Onto-Cartography: An Ontology of Machines and Media* (Edinburgh: Edinburgh University Press, 2014); and Morton, *Hyperobjects*. My all-too-brief account of Harman's thought below is based primarily on his *Tool-Being*, although the arguments can also be found repeated and developed in his later books.

9. Harman, *Tool-Being*, 2.

10. Shaviro, *Universe of Things*, 50.

11. Harman, *Tool-Being*, 16, 19, 31. Ian Bogost refers to these as "Latour litanies," and indeed he has created a "Latour litanizer" that pulls random objects from Wikipedia to generate such lists (Ian Bogost [website], http://bogost.com/writing/blog/latour_litanizer/, published December 16, 2009). Timothy Morton proposes that such random lists "are the hallmark trope of OOO" ("Here Comes Everything: The Promise of Objected-Oriented Ontology," *Qui Parle: Critical Humanities and Social Sciences* 19 [2011]: 173).

12. Morton's "Here Comes Everything" is an early manifesto; Morton's *Hyperobjects* is a more worked-out articulation of his embrace of object-oriented ontology.

13. See especially Morton, *Ecological Thought*.

14. Morton, *Hyperobjects*, 47.

15. Morton, "Here Comes Everything," 171.

16. Morton, "Here Comes Everything," 176.

17. Morton, *Ecological Thought*, 125–26.

18. Morton, *Hyperobjects*, 198. See now the further development of the affective layers of ecological awareness—guilt, shame, melancholy, play, humor, depression, sadness, longing, joy—in Morton, *Dark Ecology*, 129–58.

19. Bennett, *Vibrant Matter*, xi.

20. On this point, see Jane Bennett's critique of the strong emphasis on the isolation of objects. as opposed to their participation in systems, processes, or relations, in "Systems and Things."

21. Bennett, *Vibrant Matter*, xii–xiii.

22. Bennett, *Vibrant Matter*, 21.

23. Bennett, *Vibrant Matter*, 31.

24. Annabel Jane Wharton, *Architectural Agents: The Delusional, Abusive, Addictive Lives of Buildings* (Minneapolis: University of Minnesota Press, 2015), 185–86.

25. Schaefer, *Religious Affects*, 108.

26. The insight that the literary imagination is a crucial site for the apprehension of "the excess, the surplus value, of things" is central to the argument of Patricia Cox Miller, *The Corporeal Imagination: Signifying the Holy in Late Ancient Christianity*, Divinations: Rereading Late Ancient Religion (Philadelphia: University of Pennsylvania Press, 2009), 6.

27. For Gregory's *Life of Macrina*, I have used the edition in Gregory of Nyssa, *Vie de Sainte Macrine*, ed. Pierre Maraval, Sources Chrétiennes 178 (Paris: Cerf, 1971); translations follow Gregory of Nyssa, *Ascetical Works*, trans. Virginia Woods Callahan, Fathers of the Church (Washington, DC: Catholic University of America Press, 1967).

28. The affinities of rust with blood are intriguing, signaled on the one hand by the fact that contact with human blood makes iron rust more quickly and on the other hand by the fact that rust itself staunches wounds (Pliny, Natural History 2.41, 45).

29. Franz Joseph Dölger, "Das Anhängekreuzchen der hl. Makrina und ihr Ring mit der Kreuzpartikel: Ein Beitrag zur religiösen Volkskunde des 4. Jahrhunderts nach der Vita Macrinae des Gregor von Nyssa," *Antike und Christentum* 3 (1932): 109.

30. Jan Willem Drijvers, *Helena Augusta: The Mother of Constantine the Great and the Legend of Her Finding of the True Cross* (Leiden, Netherlands: Brill, 1992), 111.

31. Gary Vikan, "Art and Marriage in Early Byzantium," *Dumbarton Oaks Papers* 44 (1990): 147.

32. Maria Parani, "Byzantine Bridal Costume," in *Δώρημα: A Tribute to the A. G. Leventis Foundation on the Occasion of Its 20th Anniversary* (Nicosia, Cyprus: A. G. Leventis Foundation, 2000), 207.

33. Roland Barthes, *Camera Lucida: Reflections on Photography*, trans. Richard Howard (New York: Hill and Wang, 1981; French, 1980), 26.

34. Lieselotte Kötzsche-Breitenbruch, "Zum Ring von Gregor von Nyssa," in *Tesserae: Festschrift für Josef Engemann*, ed. Ernst Dassmann (Münster: Aschendorff, 1991), 291–98. As Kötzsche-Breitenbruch notes, the ring is housed in the Hélène Stathatos Collection in the National Archaeological Museum of Athens. A second, nearly identical ring was formerly in a private collection in Munich; its current location is unknown. See also Jeffrey Spier, "Some Unconventional Early Byzantine Rings," in *"Intelligible Beauty": Recent Research on Byzantine Jewelry*, ed. Chris Entwistle and Noël Adams (London: British Museum, 2010), 17, 19 n. 47, with reference to the ring in the Stathatos collection.

35. Kötzsche-Breitenbruch, "Zum Ring von Gregor von Nyssa," 291.

36. Glenn Peers, *Sacred Shock: Framing Visual Experience in Byzantium* (University Park: Pennsylvania State University Press, 2004), 34.

37. Peers, *Sacred Shock*, 134. On the "dazzle" and "incandescence" of relics, see also Miller, *Corporeal Imagination*, 62–81.

38. In exploring possible "symptoms of feeling and sensing" on the part of icons, Peers argues intriguingly that human responses of "embracing and kissing and genuflecting are also answers to their need for touch and love" ("Sense Lives of Byzantine Things," in *Knowing Bodies, Passionate Souls: Sense Perceptions in Byzantium*, ed. Susan Ashbrook Harvey and Margaret Mullett [Washington, DC: Dumbarton Oaks, 2017], 17).

39. Peers argues that some nonhuman objects "have better patience and endurance than we do, and they can teach us our dependence on them; they can show temporary abatement of the void" ("Sense Lives," 30).

40. Peers, "Object Relations," 980.

41. Following Bachelard, Miller goes so far as to speak of the imaginative transformation of sensation as a *deformation* of perception: "Deforming, imagination creates. . . . It was this kind of vision that was needed for perceiving the extra-human value in martyrs' bones that transformed them into relics" (Miller, *Corporeal Imagination*, 80).

42. As Peers notes, "The crux in such statements is always the 'as if.' We take it normally to implicate metaphorical force behind the observation. . . . But perhaps that kind of seeing was real enough in that world, after all" ("Object Relations," 978).

43. Note the slight shift in agency that accompanies the shift from an intransitive to a transitive verb form: the rocks seem to split spontaneously or through their own agency, while humans must actively split their own hearts.

44. Peers, *Sacred Shock*, 133.

45. "While at the same time the relic is thus made 'fully visible' in its power and associations, it is also unquestionably hidden from view" (Cynthia Hahn, "What Do Reliquaries Do for Relics?" *Numen* 57 [2010]: 289–90). Jaś Elsner echoes the point: "One aspect of the Christian, and especially the Byzantine, redefinition of the holy lies in its passage into a hidden space of mysterious (even mystic) secrecy whose existence was announced and access to which was enabled through material framing and visual representation generally" ("Relic, Icon, and Architecture: The Material Articulation of the Holy in East Christian Art," in *Saints and Sacred Matter: The Cult of Relics in Byzantium and Beyond*, ed. Cynthia Hahn and Holger Klein [Washington, DC: Dumbarton Oaks, 2015], 13).

46. Page duBois, *Sappho Is Burning* (Chicago: University of Chicago Press, 1995), 53.

47. As Hahn notes, reliquaries "are constantly remade" (Hahn, "What Do Reliquaries Do for Relics?" 291; Cynthia Hahn, *Strange Beauty: Issues in the Making and Meaning of Reliquaries, 400–circa 1204* [University Park: Pennsylvania State University Press, 2012], 9).

48. Jean-Louis Chrétien, *The Call and the Response*, trans. Anne A. Davenport (New York: Fordham University Press, 2004), 83.

49. Peers, *Sacred Shock*, 134.

50. Miller, *Corporeal Imagination*, 39.

51. Or better yet, "*whom* we adored and kissed": one manuscript (BrB) reads "*quem*."

52. The placard reads as follows: "*Eulogia* juglets. Jerusalem, Byzantine Period, pottery. 'They offer oil to be blessed in little flasks. When the mouth of one of the flasks touches the Wood of the Cross, the oil instantly bubbles over' (Piacenza Pilgrim)."

53. See Holt Parker, "Women and Medicine," in *A Companion to Women in the Ancient World*, ed. Sharon L. James and Sheila Dillon (Chichester, UK: Wiley Blackwell, 2012), 112.

54. Martina Bagnoli et al., *Treasures of Heaven: Saints, Relics, and Devotion in Medieval Europe* (New Haven, CT: Yale University Press, 2010), 43, catalog #23.

55. Miller, "Figuring Relics: A Poetics of Enshrinement," in *Saints and Sacred Matter: The Cult of Relics in Byzantium and Beyond*, ed. Cynthia Hahn and Holger A. Klein (Washington, DC: Dumbarton Oaks, 2015), 99.

56. Derek Krueger, "The Religion of Relics in Late Antiquity and Byzantium," in Bagnoli, *Treasures of Heaven*, 11. See also Derek Krueger, "Liturgical Time and Holy Land Reliquaries in Early Byzantium," in *Saints and Sacred Matter: The Cult of Relics in Byzantium*

and Beyond, ed. Cynthia Hahn and Holger A. Klein (Washington, DC: Dumbarton Oaks, 2015), 125.

57. Peers, *Sacred Shock*, 19.

58. Miller, *Corporeal Imagination*, 3–7.

59. It is not uniquely mine: see Glenn Peers's argument "for a deeply relational sympathy between late antique Christians and their objects" that extends "into the material world around those people" ("Object Relations," 971). Peers presses me to take the animist argument further, considering not only how humans cultivate relationships with holy objects but also how such objects might be thought to cultivate relationships with humans "to reflect and germinate their power and glory."

60. Tim Ingold, *Making: Anthropology, Archaeology, Art and Architecture* (London: Routledge, 2013), 21. With specific reference to late antiquity, see Glenn Peers, "Late Antique Making and Wonder" (forthcoming).

61. Chrétien, *Call and Response*, 85–86.

62. Chrétien, *Call and Response*, 90.

63. Peers, "Object Relations," 977.

64. Peers, "Sense Lives," 16.

65. Leslie Hill, *Maurice Blanchot and Fragmentary Writing: A Change of Epoch* (London: Continuum, 2012), 432–33.

66. Peers, "Object Relations," 975.

67. Wharton, *Architectural Agents*, xxi.

68. Catherine M. Chin, "Apostles and Aristocrats," in *Melania: Early Christianity Through the Life of One Family*, ed. Catherine M. Chin and Caroline T. Schroeder (Berkeley: University of California Press, 2017), 30.

69. To borrow a phrase well known from Hans Urs von Balthasar's study of Maximus the Confessor.

70. Amy Papalexandrou, "Text in Context: Eloquent Monuments and the Byzantine Beholder," *Word and Image* 17, no. 3 (2001): 261–62.

71. Jean-Luc Nancy, *Listening*, trans. Charlotte Mandell (New York: Fordham University Press, 2007), 34.

72. Compare the Armenian Cathedral of Saint James in Jerusalem, which houses an altar containing three rocks, said to be from Mount Sinai, Mount Tabor, and the Jordan River. The rocks are visible through a glass panel, in which three circular openings have been cut, allowing the stones to be not only seen but also touched. Visitors to the church can thus make a virtual pilgrimage to holy places they might otherwise never see.

73. Claudine Dauphin, "From Apollo and Asclepius to Christ: Pilgrimage and Healing at the Temple and Episcopal Basilica at Dor," *Liber annuus* 49 (1999): 402. On the Dor basilical complex, see also Claudine Dauphin, "On the Pilgrim's Way to the Holy City of Jerusalem," in *Archaeology and Biblical Interpretation*, ed. John R. Bartlett (London: Routledge, 1997), 145–66.

74. As Naomi Koltun-Fromm argues, the cosmogonic associations of Golgotha (which go back as early as Origen of Alexandria) emerge in the fourth century in competitive tension with parallel developments in Jewish thought regarding the rock of the Temple Mount ("Rock over Water: Pre-Historic Rocks and Primordial Waters from Creation to Salvation in Jerusalem," in *Jewish and Christian Cosmogony in Late Antiquity*, ed. Lance Jenott and Sarit Kattan Gribetz [Tübingen, Germany: Mohr Siebeck, 2013], 239–54.)

75. Joseph Leibovitch, "The Reliquary Column of Dor," *Christian News from Israel* 5, no. 1–2 (1953): 22. Leibovitch also notes the existence of a similarly inscribed rock found in the ruins of the Byzantine basilica on Mount Gerazim, a rectangular stone bearing the words "ΛΙΘΟC ΕΚ ΤΟ(Υ) ΑΓΗΙΟΥ ΚΡΑΝΙΟΥ" above a round cavity (23).

76. Dauphin, "From Apollo and Asclepius to Christ," 403–5.

77. Dauphin, "From Apollo and Asclepius to Christ," 404.

78. Dauphin, "From Apollo and Asclepius to Christ," 406. Note that remains of two similar columns still lie on the site of the Dor church, though I have not been able to determine if they are exact matches for ours.

79. The inscription is graffito-like in its simplicity, its relative crudeness, and its appearance on the surface of what was at least at some point a part of the supporting structure of a building, all of which lends it a certain informality and intimacy. However, the scale of the lettering and (even more so) the embedded reliquary and plaque suggest a more "official" engraving; note too that the inscription has not attracted other graffiti, as one would expect, were this carving an act of popular devotion. On the materiality of graffiti in Christian holy places, see Ann Marie Yasin, "Prayers on Site: The Materiality of Devotional Graffiti and the Production of Early Christian Sacred Space," in *Viewing Inscriptions in the Late Antique and Medieval World*, ed. Antony Eastmond (New York: Cambridge University Press, 2015), 36–60. We are reminded that the inscriptional adornment of churches was a collective process and did not always fall into clear-cut categories of authorized, professional engravings and unauthorized, informal scratchings. For a discussion of inscriptions on church columns that seem to blur the boundaries between official and unofficial, see Stefanos Alexopoulos, "When a Column Speaks: The Liturgy of the Christian Parthenon," *Dumbarton Oaks Papers* 69 (2015): 159–78.

80. Dauphin, "From Apollo and Asclepius to Christ," 403.

81. Dauphin, "From Apollo and Asclepius to Christ," 406.

82. Dauphin, "From Apollo and Asclepius to Christ," 397–98, 400.

83. Cf. Peers, who argues for "eradicating the use of metaphor and representation as explanatory terms within Byzantine materiality," in favor of a theory of animism more native to Byzantine thought itself ("Byzantine Things," 41).

84. Tomas Lehmann suggests that the clerestory of Paulinus's basilica was set a bit higher than usual, with the result that "the illumination of the Nolan apse and its vault mosaic through the well-lit nave was thereby clearly limited"; this would have made the light emanating from the windowed apse, reflecting off the mosaic vault and marble walls and floor, all the more striking and eye catching, he proposes (*Paulinus Nolanus und die Basilica Nova in Cimitile/Nola* [Wiesbaden, Germany: Reichert, 2004], 247).

85. Wikimedia Commons, Ravenna (Italy), Sant' Apollinare in Classe, cupola, https://upload.wikimedia.org/wikipedia/commons/4/42/Ravenna_BW_4.JPG.

86. 360cities.net, Liguria, Italy, Albenga Baptistery, inside view, http://www.360cities.net/image/albenga-baptistery-liguria-italy. Robin Jensen notes, "Given their number, the doves in the mosaic probably refer to the apostles, but they may also symbolize the newly baptized, who have been made newly innocent" (*Baptismal Imagery in Early Christianity: Ritual, Visual, and Theological Dimensions* [Grand Rapids, MI: Baker Academic], 119).

87. Wikimedia Commons, Ravenna (Italy), Arian Baptistry, ceiling mosaic, https://upload.wikimedia.org/wikipedia/commons/6/60/Arian_Baptistry_ceiling_mosaic_-_Ravenna.jpg. On the gushing dove, see Jensen, *Baptismal Imagery*, 120.

88. Rudolf Carel Goldschmidt, *Paulinus' Churches at Nola: Texts, Translations, and Commentary* (Amsterdam: N.V. Noord-Hollandische Uitgevers Maatschappij, 1940), 98.

89. proFBonomo (blog), "Santi Apostoli Pietro e Paolo," June 29, 2013, http://francescobonomo.blogspot.co.il/2013_06_01_archive.html.

90. Other crosses in the church also create relational networks, such as the wreathed cross pictured above the entrance and the two scarlet crosses pictured on either side of the entrance, all inscribed with verses (Ep. 32.12, 14).

91. On the archaeological and literary evidence for this courtyard, see Lehmann, *Paulinus Nolanus*, 243–44.

92. On the *cantharus* and other kinds of vessels named, especially as these relate to archaeological discoveries on the site, see Annewies van den Hoek and John Joseph Hermann Jr., *Pottery, Pavements, and Paradise: Iconographic and Textual Studies on Late Antiquity* (Leiden, Netherlands: Brill, 2013), 9–63.

93. Gaston Bachelard, *The Poetics of Space*, trans. Maria Jolas (Boston: Beacon, 1994; French, 1958), 51.

94. The authoritative work on the building and mosaics of Santa Costanza is Jürgen J. Rasch and Achim Arbeiter, *Das Mausoleum der Constantina in Rom* (Mainz: Philipp von Zabern, 2007).

95. An inscription testifies that the basilica was commissioned and paid for by Constantina herself; construction was likely begun around 337. The mausoleum may have been begun a few years later, in the 340s, since there are archaeological indications that the original plan called for a much smaller, triconch structure. David Stanley, whose excavations uncovered the foundations of the triconch annex, proposes a significantly later date for the current structure ("New Discoveries at Santa Costanza," *Dumbarton Oaks Papers* 28 [1994]: 257–61). However, Jürgen Rasch argues that the decision to expand the mausoleum occurred before the triconch structure was ever built and that the mausoleum should not be dated later than the mid-fourth century (Rasch and Arbeiter, *Das Mausoleum*, 89).

96. The division of the ambulatory vault into twelve fields divides the space into visually manageable portions (Rasch and Arbeiter, *Das Mausoleum*, 192), yet glimpses of adjacent fields pull the eye along.

97. See, e.g., Roger Ling, *Ancient Mosaics* (Princeton, NJ: Princeton University Press, 1998), 106; Rasch and Arbeiter, *Das Mausoleum*, 104, 230.

98. Susan Power Bratton, *Environmental Values in Christian Art* (Albany: State University of New York, 2008), 17.

99. Arbeiter refers to a "legible increase of thematic and aesthetic demand from the entrance to the end point," with the Dionysian fields marking "clear intermediate high points" (Rasch and Arbeiter, *Das Mausoleum*, 153, cf. 230).

100. Arbeiter suggests that the decline in the forcefulness of the mosaics in the second to last field is intentional, setting off the final field by way of contrast (Rasch and Arbeiter, *Das Mausoleum*, 153, 230).

101. Rina Talgam, *Mosaics of Faith: Floors of Pagans, Jews, Samaritans, Christians, and Muslims in the Holy Land* (Jerusalem and University Park: Yad Ben-Zvi Press and Pennsylvania State University Press, 2014), 180.

102. Crosses on floors inspired particular opposition, yet Middle Eastern Christians (among others) continued to inscribe them on church pavements: see Glenn Peers, "Crosses' Work Underfoot: Christian Spolia in the Late Ancient Mosque at Shivta in the Negev Desert (Israel)," *Eastern Christian Art* 8 (2011): 109–13.

103. For an overview of the ecclesiastical complex at Petra, see Patricia Maynor Bikai, "The Churches of Byzantine Petra," *Near Eastern Archaeology* 65, no. 4 (2002): 271–76. On the dating of the mosaics, see Tomasz Waliszewski, "Mosaics," in *The Petra Church* (Amman, Jordan: American Center of Oriental Research, 2001), 244, 259, 256. The American Center of Oriental Research in Amman website provides excellent access to color photographs of the mosaics (https://www.acorjordan.org/product/petra-church-north-aisle/ and https://www.acorjordan.org/product/petra-church-south-aisle/, accessed January 21, 2018).

104. Jacques Derrida, "Typewriter Ribbon: Limited Ink (2)," in *Without Alibi*, ed. and trans. Peggy Kamuf (Stanford, CA: Stanford University Press, 2002), 128.

105. As Waliszewski notes, "Early Christian art contains representations of the Evangelists or saints holding the Gospels in the left hand, which is covered with the edge of their robes because of its holiness. . . . Representations of Christ in majesty with a book are similar" (Waliszewski, "Mosaics," 254).

106. See Talgam, *Mosaics of Faith*, 210: "Since Wisdom is but a synonym for philosophy, . . . it is possible that this was an attempt to annex the realm of science and knowledge to the Church. The incorporation of personifications derived from Classical art did not necessarily create the secularization of the sacred; on the contrary, it resulted in the creation of a composition of greater significance."

107. As Talgam notes, "The *Physiologus* refers to the eagle as a symbol of renewal," and in some contexts, eagles seem to have christological associations (*Mosaics of Faith*, 201).

108. Byzantine mosaics typically represent Christ or ecclesiastical figures holding the gospels, and perhaps that is the most likely interpretation here as well. In the context, however, it is intriguing to imagine that *sophia* holds a Genesis codex, or even a cosmological and astrological work like the late ancient Syriac calendologion known as the *Treatise of Shem* (see Meir Bar-Ilan, "The Hebrew *Book of Creation* and the Syriac *Treatise of Shem*," *ARAM* 24 [2012]: 203–18).

109. Hill, *Maurice Blanchot*, 432–33.

110. Waliszewski, "Mosaics," 219, 221.

111. Bennett, *Vibrant Matter*, 31.

112. Wharton, *Architectural Agents*, xxi.

113. Cf. Bachelard, *Poetics of Space*, 51.

114. Barthes, *Camera Lucida*, 20.

115. Bailly, *Animal Side*, 16.

116. Betcher, "Crip/Tography," 101.

117. Morton, *Ecological Thought*, 98.

118. J. L. Austin, *How to Do Things with Words*, 2nd ed. (Cambridge, MA: Harvard University Press, 1975).

119. For Basil's *Hexameron*, I have used the edition in Basil of Caesarea, *Homélies sur l'Hexaéméron*, Sources Chrétiennes 26 (Paris: Cerf, 1968); translations are my own.

120. For a complex (and explicitly post-Austinian) thinking-through of the relation of "*both* what is happening (we call that an event) *and* the calculable programming of an automatic repetition (we call that a machine)," see Derrida, "Typewriter Ribbon," 71.

121. On the place and time of Basil's delivery of these sermons, see Emmanuel Amand de Mendieta, "Les neuf homélies de Basile de Césarée sur l'Hexaéméron," *Byzantion* 48 (1978): 352, 366–67.

122. The same might be said for liturgy. Mika Ahuvia considers the resonance of the figure of the abyss in late ancient liturgical poetry and art, for example ("Darkness upon

the Abyss: Depicting Cosmogony in Late Antiquity," in *Jewish and Christian Cosmogony in Late Antiquity*, ed. Lance Jenott and Sarit Kattan Gribetz [Tübingen, Germany: Mohr Siebeck, 2013], 255–70).

123. Here I follow the translation and interpretation of J. C. M. van Winden, "'An Appropriate Beginning': The Opening Passage of Saint Basil's *In Hexaemeron*," *Revue de l'histoire des religions* 203 (1983): 307–11.

124. Note that Basil's initial nine hexameral sermons were subsequently supplemented by two sermons dealing with the creation of humanity; the relation of those sermons to the Basilian corpus is the subject of scholarly debate. Even if not precisely the author (Amand de Mendieta, "Les neuf homélies," 339), Basil may nonetheless have been the source of their ideas (Philip Rousseau, "Human Nature and Its Material Setting in Basil of Caesarea's Sermons on the Creation," *Heythrop Journal* 49 [2008]: 222). What interests me, however, is the way that the original published compilation of nine sermons reads as a whole.

125. Virginia Burrus, Mark Jordan, and Karmen MacKendrick, *Seducing Augustine: Bodies, Desires, Confessions* (New York: Fordham University Press, 2010), 69.

126. Waliszewski, "Mosaics," 249.

127. Chrétien, *Call and Response*, 31.

128. Waliszewski, "Mosaics," 244–59.

129. Chrétien, *Call and Response*, 1.

130. Miller, "Adam, Eve, and the Elephants," 262.

131. I have certainly downplayed the hierarchical aspect. For a more critical engagement of Basil's assumptions about human superiority, see Miller, "Adam, Eve, and the Elephants," 255–57.

132. See Waliszewski, "Mosaics," 231–32.

133. Alternately, "Adam's right hand . . . may point to the roaring lion in wonder" (Miller, "Adam, Eve, and the Elephants," 255).

134. Glenn Peers, "Adam's Anthropocene," *Postmedieval: A Journal of Medieval Cultural Studies* (2016): 5.

135. Chrétien, *Call and Response*, 21.

136. Suspicion of ecclesiastical music was to have an afterlife even in the hymn-loving east; see Stig Simeon Froyshov, "La réticence à l'hymnographie chez des anachorètes de l'Égypte et du Sinaï du 6e au 8e siècles," in *L'Hymnographie: Conférences Saint-Serge XLVIe Semaine d'études liturgiques, Paris 1999*, ed. A. M. Triacca and A. Pistoia (Rome: Edizioni Liturgiche, 2000), 229–45.

137. Nancy, *Listening*, 13.

138. The phrase "undulating temporality" is inspired by Morton, *Hyperobjects*, 55.

139. On liturgical psalmody in Augustine's church, see James W. McKinnon, "Liturgical Psalmody in the Sermons of St. Augustine: An Introduction," in *The Study of Medieval Chant: Paths and Bridges, East and West, in Honor of Kenneth Levy*, ed. Peter Jeffery (Cambridge: Boydell, 2001), 7–24.

140. Karmen MacKendrick, *The Matter of Voice: Sensual Soundings* (New York: Fordham University Press, 2016), 116.

141. John Peter Kenney, *The Mysticism of Saint Augustine: Rereading the* Confessions (New York: Routledge, 2005), 128.

142. Kenney, *Mysticism of Saint Augustine*, 53.

143. Kenney, *Mysticism of Saint Augustine*, 108.

144. MacKendrick, *Matter of Voice*, 116.

145. For Augustine's *City of God*, I have used the edition in Augustine, *City of God*, trans. William M. Green, Loeb Classical Library 417 (Cambridge, MA: Harvard University Press, 1972); translations are mine.

146. Margaret R. Miles, "'*Facie ad faciem*': Visuality, Desire, and the Discourse of the Other," *Journal of Religion* 87, no. 1 (2007): 57.

147. Morton, *Ecological Thought*, 70.

148. Derrida, "Animal That Therefore I Am," 21.

149. Bailly, *Animal Side*, 15.

150. Bailly, *Animal Side*, 5.

151. Chrétien, *Call and Response*, 85, 93.

152. Ian Hodder, *Entangled: An Archaeology of the Relationships Between Humans and Things* (Chichester, UK: Wiley-Blackwell, 2012), 41.

153. Bennett, *Vibrant Matter*, 61.

154. Chrétien, *Call and Response*, 131.

155. For Dionysius's works, I have used the editions in *Corpus Dionysiacum I and II*, ed. Beate Regina Suchla, Günter Heil, and Adolf Martin Ritter (Berlin: De Gruyter, 1990–91); translations generally follow John H. Parker, trans., *The Complete Works of Dionysius the Aereopagite* (London: James Parker, 1897–99).

156. This identification of *pronoia*, eros, and the one god is apparently original with Dionysius. See John M. Rist, "A Note on Eros and Agape in Pseudo-Dionysius," *Vigiliae Christianae* 20 (1966): 239–40; Cornelia J. de Vogel, "Greek Cosmic Love and the Christian Love of God: Boethius, Dionysius the Areopagite and the Author of the Fourth Gospel," *Vigiliae Christianae* 35 (1981): 70–73; and J. Warren Smith, "Divine Ecstasy and Divine Simplicity: The Eros Motif in Pseudo-Dionysius's Soteriology," *Pro Ecclesia* 21, no. 2 (2012): 212–14.

157. Eric Justin David Perl, "Symbol, Sacrament, and Hierarchy in Saint Dionysios the Areopagite," *Greek Orthodoxy Theological Review* 39, no. 3/4 (1994): 324.

158. Rist, "Eros and Agape," 243.

159. Sean Braune, "How to Analyze Texts That Were Burned, Lost, Fragmented, or Never Written," *Symplokē* 21, no. 1–2 (2013): 248.

160. Sebastian Brock and Brian Fitzgerald, trans., *Two Early Lives of Severos, Patriarch of Antioch* (Liverpool: Liverpool University Press, 2013), 126–27.

161. Bryan D. Spinks, *The Sanctus in the Eucharistic Prayer* (Cambridge: Cambridge University Press, 1991), 117.

162. Perl, "Symbol, Sacrament, and Hierarchy," 321.

163. Perl, "Symbol, Sacrament, and Hierarchy," 319.

164. Charles M. Stang, *Apophasis and Pseudonymity in Dionysius the Areopagite: "No Longer I"* (Oxford: Oxford University Press, 2012), 83–85.

165. Stang, *Apophasis and Pseudonymity*, 196.

166. Stang, *Apophasis and Pseudonymity*, 89–98.

167. Mary-Jane Rubenstein, "Dionysius, Derrida, and the Critique of 'Ontotheology,'" *Modern Theology* 24, no. 4 (2008): 737.

168. Charles M. Stang, "The Beginning and End of All Hierarchy," in *The Open Body: Essays in Anglican Ecclesiology*, ed. Zachary Guiliano and Charles M. Stang (New York: Peter Lang, 2012), 107–8.

169. Rubenstein, "Dionysius, Derrida, and the Critique of 'Ontotheology,'" 737.

170. An obvious problem with Dionysius's thought is the naturalization of human (more specifically, ecclesiastical) hierarchies; see Stang, "Beginning and End of All Hierarchy." But is all hierarchical order bad? "Morality, like the symbolic, emerges within—not beyond—the human. Projecting our morality, which rightfully privileges equality, on a relational landscape composed in part of nested and unidirectional associations of a logical and ontological, but not a moral, nature is a form of anthropocentric narcissism that renders us blind to some of the properties of that world beyond the human" (Eduardo Kohn, *How Forests Think: Toward an Anthropology Beyond the Human* [Berkeley: University of California Press, 2013], 19).

171. "Dionysios is one of the few Fathers . . . to affirm unequivocally that we *know God by sense perception*. . . . Dionysios specially insists that we must literally see, smell, touch God with our bodily senses" (emphasis in original) (Perl, "Symbol, Sacrament, and Hierarchy," 327).

172. Mary-Jane Rubenstein, "The Lure of Pan(en)theism: Difference and Desire in *Divine Enticement*," *Theology and Sexuality* 18, no. 2 (2012): 117.

173. Kohn, *How Forests Think*, 22.

174. Morton, *Ecological Thought*, 96.

175. Nancy, *Listening*, 7.

176. Bogost, *Alien Phenomenology*, 38.

177. Bogost, *Alien Phenomenology*, 38.

178. Graham Harman, *Prince of Networks: Bruno Latour and Metaphysics* (Melbourne: Re.press, 2009), 102.

179. Bennett, *Vibrant Matter*, ix.

180. Morton, *Ecological Thought*, 104.

181. Morton, *Ecological Thought*, 134.

182. Morton, *Ecological Thought*, 9.

183. Morton, *Ecological Thought*, 127.

184. Christie, *Blue Sapphire of the Mind*.

185. See Bogost, *Alien Phenomenology*, 38–39, 94–96; and Morton, "Here Comes Everything," 173–74.

186. Bennett, *Vibrant Matter*, 122.

187. Chrétien, *Call and Response*, 83.

188. Jean-Luc Nancy, *The Inoperative Community*, ed. Peter Connor, trans. Peter Connor, et al., Theory and History of Literature, Vol. 76 (Minneapolis: University of Minnesota Press, 1991), 149.

189. Nancy, *Inoperative Community*, 137.

190. Nancy, *Inoperative Community*, 143.

191. Nancy, *Inoperative Community*, 150.

Epilogue. Worm Stories

1. Charles Darwin, *The Formation of Vegetable Mould, Through the Action of Worms, with Observations on Their Habits* (New York: D. Appleton & Company, 1882), 305.

2. Darwin, *Formation of Vegetable Mould*, 305, 308–9, 313.

3. Bennett, *Vibrant Matter*, 108.

4. Bennett, *Vibrant Matter*, 94.

5. Adam Phillips, *Darwin's Worms: On Life Stories and Death Stories* (New York: Basic, 2000), 58.

6. Phillips, *Darwin's Worms*, 14, 12.

7. Ruaro, "Resurrection: The Hope of Worms."

8. Daniel A. Bertrand, "Le Christ comme ver; a propos du Psaume 22(21)7," *Cahiers de Biblia patristica* 4 (1994): 222.

9. Ruaro, "Resurrection: The Hope of Worms," 121.

10. Smithsonian National Museum of Natural History, Explore Our Collections, "Giant Tube Worm: *Riftia pachyptila*," accessed July 4, 2017, https://naturalhistory.si.edu/onehundredyears/featured_objects/Riftia.html.

11. Phillips, *Darwin's Worms*, 8.

12. Warr, *Artist's Body*, 101.

13. Iain S. Whitaker et al., "Larval Therapy from Antiquity to the Present Day: Mechanisms of Action, Clinical Applications and Future Potential," *Postgraduate Medical Journal* 83 (2007): 409–13. See Pliny, *Natural History* 30.39, on the curative properties of worms.

14. Enrica Ruaro, "God and the Worm: The Twofold Otherness in Pseudo-Dionysius's Theory of Dissimilar Images," *American Catholic Philosophical Quarterly* 82, no. 4 (2008): 581–92.

Bibliography

Ahmed, Sara. "Happy Objects." In *The Affect Theory Reader*, edited by Melissa Gregg and Gregory J. Seigworth, 29–51. Durham, NC: Duke University Press, 2010.

Ahuvia, Mika. "Darkness upon the Abyss: Depicting Cosmogony in Late Antiquity." In *Jewish and Christian Cosmogony in Late Antiquity*, edited by Lance Jenott and Sarit Kattan Gribetz, 255–70. Tübingen, Germany: Mohr Siebeck, 2013.

Alexopoulos, Stefanos. "When a Column Speaks: The Liturgy of the Christian Parthenon." *Dumbarton Oaks Papers* 69 (2015): 159–78.

Ali, Amanat, Mostafa Waly, M. Mohamed Essa, and Sankar Devarajan. "Nutritional and Medicinal Value of Date Fruit." In *Dates: Production, Processing, Food, and Medicinal Values*, edited by A. Manickavasagan, M. Mohamed Essa, and E. Sukumar, 361–75. Boca Raton, FL: CRC Press, 2012.

Allison, Dorothy. *Skin: Talking About Sex, Class, and Literature*. Ithaca, NY: Firebrand, 1994.

Amand de Mendieta, Emmanuel. "Les neuf homélies de Basile de Césarée sur l'Hexaéméron." *Byzantion* 48 (1978): 337–68.

Anatolios, Khaled. *Athanasius: The Coherence of His Thought*. London: Routledge, 1998.

Arruzza, Cinzia. "Plato's World-Maker in Origen's *Contra Celsum*." *Horizons* 3, no. 1 (2012): 61–80.

Athanasius. Contra gentes *and* De incarnatione. Edited and translated by Robert W. Thomson. Oxford: Clarendon, 1971.

———. *The Life of Antony and the Letter to Marcellinus*. Translated and introduced by Robert C. Gregg. The Classics of Western Spirituality. New York: Paulist Press, 1980.

———. *Opera omnia, tomus quartus*. Patrologia Graeca 28, 1887.

———. *Vie d'Antoine*. Edited by G. J. M. Bartelink. Sources Chrétiennes 400. Paris: Cerf, 1994.

Augustine. *City of God*. Translated by William M. Green. Loeb Classical Library 417. Cambridge, MA: Harvard University Press, 1972.

———. *Confessions*. Translated by Carolyn J.-B. Hammond. Loeb Classical Library 26 and 27. Cambridge, MA: Harvard University Press, 2014–16.

———. *Opera omnia, tomus tertius*. Patrologia Latina 34, 1841.

———. *Saint Augustine: On Genesis*. Translated and edited by Edmund Hill. Hyde Park, NY: New City Press, 2002.

Austin, J. L. *How to Do Things with Words*. 2nd ed. Cambridge, MA: Harvard University Press, 1975.

Bachelard, Gaston. *The Poetics of Space*. Translated by Maria Jolas. Boston: Beacon, 1994; French, 1958.

Bagemihl, Bruce. *Biological Exuberance: Animal Homosexuality and Natural Diversity*. New York: St. Martin's, 1999.

Bailly, Jean-Christophe. *The Animal Side*. Translated by Catherine Porter. New York: Fordham University Press, 2011.

Bar-Ilan, Meir. "The Hebrew *Book of Creation* and the Syriac *Treatise of Shem*." *ARAM* 24 (2012): 203–18.

Barnes, Timothy D. *Athanasius and Constantius: Theology and Politics in the Constantinian Empire*. Cambridge, MA: Harvard University Press, 1993.

Barthes, Roland. *Camera Lucida: Reflections on Photography*. Translated by Richard Howard. New York: Hill and Wang, 1981; French, 1980.

Basil of Caesarea. *Homélies sur l'hexaéméron*. Sources Chrétiennes 26. Paris: Cerf, 1968.

Bauman, Whitney A. "Disability Studies, Queer Theory, and the New Materialism." *Worldviews* 19, no. 1 (2015): 69–73.

———. *Religion and Ecology: Developing a Planetary Ethic*. New York: Columbia University Press, 2014.

———. "What's Left (Out) of the Lynn White Narrative?" In *Religion and Ecological Crisis: The "Lynn White Thesis" at Fifty*, edited by Todd LeVasseur and Anna Peterson, 165–77. New York: Routledge, 2017.

Bennett, Jane. "Systems and Things: A Response to Graham Harman and Timothy Morton." *New Literary History* 43 (2012): 225–33.

———. *Vibrant Matter: A Political Ecology of Things*. Durham, NC: Duke University Press, 2010.

Bernstein, Neil W. "*Locus Amoenus* and *Locus Horridus* in Ovid's *Metamorphoses*." *Wenshan Review of Literature and Culture* 5 (2011): 67–98.

Bertrand, Daniel A. "Le Christ comme ver; a propos du Psaume 22(21)7." *Cahiers de Biblia patristica* 4 (1994): 221–34.

Betcher, Sharon V. "Crip/Tography: Disability Theology in the Ruins of God." *Journal for Cultural and Religious Theory* 14, no. 3 (2015): 98–115.

———. "The Picture of Health: 'Nature' at the Intersection of Disability, Religion, and Ecology." *Worldviews* 19, no. 1 (2015): 9–33.

———. *Spirit and the Obligation of Social Flesh: A Secular Theology for the Global City*. New York: Fordham University Press, 2014.

Bikai, Patricia Maynor. "The Churches of Byzantine Petra." *Near Eastern Archaeology* 65, no. 4 (2002): 271–76.

Black, Carolyn. "Gina Pane: Arnolfini, Bristol, 23.02.02 to 13.04.02." *Decide Magazine* 2 (2002): 11.

Boesel, Chris, and Catherine Keller. "Introduction." In *Apophatic Bodies: Negative Theology, Incarnation, and Relationality*, edited by Chris Boesel and Catherine Keller, 1–21. New York: Fordham University Press, 2010.

Bogost, Ian. *Alien Phenomenology, or What It's Like to Be a Thing*. Minneapolis: University of Minnesota Press, 2012.

Boon, Marcus, Eric Cazdyn, and Timothy Morton. *Nothing: Three Inquiries in Buddhism*. Chicago: University of Chicago Press, 2015.

Brakke, David. "Athanasius." In *Augustine Through the Ages: An Encyclopedia*, 77–79. Grand Rapids, MI: Eerdmans, 1999.

———. *Demons and the Making of the Monk: Spiritual Combat in Early Christianity.* Cambridge, MA: Harvard University Press, 2006.

Bratton, Susan Power. *Environmental Values in Christian Art.* Albany: State University of New York, 2008.

Braune, Sean. "How to Analyze Texts That Were Burned, Lost, Fragmented, or Never Written." *Symplokē* 21, no. 1–2 (2013): 239–55.

Brock, Sebastian, and Brian Fitzgerald, trans. *Two Early Lives of Severos, Patriarch of Antioch.* Liverpool: Liverpool University Press, 2013.

Brown, Bill. "Thing Theory." *Critical Inquiry* 28, no. 1 (2001): 1–22.

Bryant, Levi R. *The Democracy of Objects.* Ann Arbor, MI: Open Humanities Press, 2011.

———. *Onto-Cartography: An Ontology of Machines and Media.* Edinburgh: Edinburgh University Press, 2014.

Buell, Lawrence. *The Future of Environmental Criticism: Environmental Crisis and Literary Imagination.* Oxford: Blackwell, 2005.

Burrus, Virginia. "Bodies, Desires, Confessions: Shame in Plotinus, Antony, and Augustine." In *Shame Between Penance and Punishment*, edited by Bénédicte Sère and Jörg Wettlaufer, 23–48. Florence: SISMEL-Edizioni del Galluzzo, 2013.

———. *The Sex Lives of Saints: An Erotics of Ancient Hagiography.* Divinations: Rereading Late Ancient Religion. Philadelphia: University of Pennsylvania Press, 2004.

———. "Wyschogrod's Hand: Saints, Animality, and the Labor of Love." *Philosophy Today* 55, no. 4 (2011): 412–21.

Burrus, Virginia, Mark Jordan, and Karmen MacKendrick. *Seducing Augustine: Bodies, Desires, Confessions.* New York: Fordham University Press, 2010.

Calarco, Matthew. *Thinking Through Animals: Identity, Difference, Indistinction.* Stanford, CA: Stanford University Press, 2015.

Caputo, John D. *The Insistence of God: A Theology of Perhaps.* Bloomington: Indiana University Press, 2013.

———. *The Prayers and Tears of Jacques Derrida: Religion Without Religion.* Bloomington: Indiana University Press, 1997.

———. *The Weakness of God: A Theology of the Event.* Bloomington: Indiana University Press, 2006.

Castelli, Elizabeth A. "Life of Syncletica." In *Ascetic Behavior in Greco-Roman Antiquity*, edited by Vincent L Wimbush, 256–311. Minneapolis: Fortress, 1990.

Chadwick, Henry. "Origen, Celsus and the Stoa." *Journal of Theological Studies* 48 (1947): 34–49.

Chapple, Christopher Key. "Lynn White Jr. and India: Romance? Reality?" In *Religion and Ecological Crisis: The "Lynn White Thesis" at Fifty*, edited by Todd LeVasseur and Anna Peterson, 110–20. New York: Routledge, 2017.

Chen, Mel Y. *Animacies: Bioplitics, Racial Mattering, and Queer Affect.* Durham, NC: Duke University Press, 2012.

———. "The Reproduction in/of Disability and Environment." *Worldviews* 19, no. 1 (2015): 79–82.

Chin, Catherine M. "Apostles and Aristocrats." In *Melania: Early Christianity Through the Life of One Family.* Edited by Catherine M. Chin and Caroline T. Schroeder, 19–33. Berkeley: University of California Press, 2017.

Chisholm, Dianne. "Biophilia, Creative Involution, and the Ecological Future of Queer Desire." In *Queer Ecologies: Sex, Nature, Politics, Desire*, edited by Catriona Mortimer-

Sandilands and Bruce Erickson, 359–81. Bloomington: Indiana University Press, 2010.

Chrétien, Jean-Louis. *The Call and the Response*. Translated by Anne A. Davenport. New York: Fordham University Press, 2004.

Christie, Douglas E. *The Blue Sapphire of the Mind: Notes for a Contemplative Ecology*. New York: Oxford University Press, 2013.

Clough, David. "All God's Creatures: Reading Genesis on Human and Non-Human Animals." In *Reading Genesis After Darwin*, edited by Stephen C. Barton and David Wilkinson, 145–61. Oxford: Oxford University Press, 2009.

Colebrook, Claire. "We Have Always Been Post-Anthropocene: The Anthropocene Counter-Factual." Academia.edu, 2014. https://www.academia.edu/12757260/We_Have_Always_Been_Post-Anthropocene.

Coole, Diana, and Samantha Frost. "Introducing the New Materialisms." In *New Materialisms: Ontology, Agency, and Politics*, edited by Diana Coole and Samantha Frost, 1–43. Durham, NC: Duke University Press, 2010.

Coon, Lynda L. *Sacred Fictions: Holy Women and Hagiography in Late Antiquity*. Middle Ages Series. Philadelphia: University of Pennsylvania Press, 1997.

Curtius, Ernst Robert. *European Literature and the Latin Middle Ages*. Translated by Willard R. Trask. Princeton, NJ: Princeton University Press, 1953; German original, 1948.

Darwin, Charles. *The Formation of Vegetable Mould, Through the Action of Worms, with Observations on Their Habits*. New York: D. Appleton, 1882.

Dauphin, Claudine. "From Apollo and Asclepius to Christ: Pilgrimage and Healing at the Temple and Episcopal Basilica at Dor." *Liber annuus* 49 (1999): 397–430.

———. "On the Pilgrim's Way to the Holy City of Jerusalem." In *Archaeology and Biblical Interpretation*, edited by John R. Bartlett, 145–66. London: Routledge, 1997.

de Vogel, Cornelia J. "Greek Cosmic Love and the Christian Love of God: Boethius, Dionysius the Areopagite and the Author of the Fourth Gospel." *Vigiliae Christianae* 35 (1981): 57–81.

Deleuze, Gilles, and Felix Guattari. *A Thousand Plateaus: Capitalism and Schizophrenia*. Translated by Brian Massumi. Minneapolis: University of Minnesota Press, 1987.

Derrida, Jacques. *The Animal That Therefore I Am*. Edited by Marie-Louise Mallet. Translated by David Wills. New York: Fordham University Press, 2008.

———. "How to Avoid Speaking: Denials." In *Derrida and Negative Theology*, edited by Harold Coward and Toby Foshay, 73–142. Albany: State University of New York Press, 1992.

———. "Khōra." In *On the Name*, translated by David Wood, John P. Leavey, and Ian McLeod, 87–127. Stanford, CA: Stanford University Press, 1995.

———. "Typewriter Ribbon: Limited Ink (2)." In *Without Alibi*, edited and translated by Peggy Kamuf, 71–160. Stanford, CA: Stanford University Press, 2002.

Digeser, Elizabeth DePalma. "Origen on the *Limes*: Rhetoric and the Polarization of Identity in the Late Third Century." In *The Rhetoric of Power in Late Antiquity: Religion and Politics in Byzantium, Europe and the Early Islamic World*, edited by Robert M. Frakes, Elizabeth DePalma Digeser, and Justin Stephens, 197–218. London: Tauris Academic Studies, 2010.

———. *A Threat to Public Piety: Christians, Platonists, and the Great Persecution*. Ithaca, NY: Cornell University Press, 2012.

Dillon, John. "The *Timaeus* in the Old Academy." In *Plato's* Timaeus *as Cultural Icon*, edited by Gretchen J. Reydams-Schils, 80–94. Notre Dame, IN: University of Notre Dame Press, 2003.

Dionysius the Aereopagite. *The Complete Works of Dionysius the Aereopagite*. Translated by John H. Parker. London: James Parker, 1897–99.

———. *Corpus Dionysiacum I and II*. Edited by Beate Regina Suchla, Günter Heil, and Adolf Martin Ritter. Berlin: De Gruyter, 1990–91.

Dölger, Franz Joseph. "Das Anhängekreuzchen der hl. Makrina und ihr Ring mit der Kreuzpartikel: Ein Beitrag zur religiösen Volkskunde des 4. Jahrhunderts nach der Vita Macrinae des Gregor von Nyssa." *Antike und Christentum* 3 (1932): 81–116.

Doran, Robert. *The Lives of Simeon Stylites*. Kalamazoo, MI: Cistercian Publications, 1992.

Dörrie, Heinrich. *Von Platon zum Platonismus: Ein Bruch in der Überlieferung und seine Überwindung*. Rheinisch-Westfälische Akademie der Wissenschaft Vorträge. Opladen, Germany: Westdeutscher Verlag, 1976.

Drijvers, Jan Willem. *Helena Augusta: The Mother of Constantine the Great and the Legend of Her Finding of the True Cross*. Leiden, Netherlands: Brill, 1992.

duBois, Page. *Sappho Is Burning*. Chicago: University of Chicago Press, 1995.

Edwards, Mark J. "Ammonius, Teacher of Origen." *Journal of Ecclesiastical History* 44 (1993): 1–13.

———. "*Locus Horridus* and *Locus Amoenus*." In Homo Viator*: Classical Essays for John Bramble*, edited by Michael Whitby, Philip Hardie, and Mary Whitby, 267–76. Bristol, UK: Bristol Classical Press, 1987.

Elsner, Jaś. "Relic, Icon, and Architecture: The Material Articulation of the Holy in East Christian Art." In *Saints and Sacred Matter: The Cult of Relics in Byzantium and Beyond*, edited by Cynthia Hahn and Holger Klein, 13–39. Washington, DC: Dumbarton Oaks, 2015.

Erickson, Bruce, and Catriona Mortimer-Sandilands "Introduction: A Genealogy of Queer Ecologies." In *Queer Ecologies: Sex, Nature, Politics, Desire*, edited by Bruce Erickson and Catriona Mortimer-Sandilands, 1–47. Bloomington: Indiana University Press, 2010.

Finkelberg, Aryeh. "Plato's Method in *Timaeus*." *American Journal of Philology* 117 (1996): 391–409.

Fischel, Henry A. "The Transformation of Wisdom in the World of Midrash." In *Aspects of Wisdom in Judaism and Early Christianity*, edited by Robert L. Wilken, 67–101. Notre Dame, IN: University of Notre Dame Press, 1975.

Freeman, Elizabeth. "Time Binds, or, Erotohistoriography." *Social Text* 23, no. 3–4, 84–85 (2005): 57–68.

———. *Time Binds: Queer Temporalities, Queer Histories*. Durham, NC: Duke University Press, 2010.

Froyshov, Stig Simeon. "La réticence à l'hymnographie chez des anachorètes de l'Égypte et du Sinaï du 6e au 8e siècles." In *L'Hymnographie: Conférences Saint-Serge XLVIe Semaine d'études liturgiques, Paris 1999*, edited by A. M. Triacca and A. Pistoia, 229–45. Rome: Edizioni Liturgiche, 2000.

Furstenberg, Yair. "The Rabbinic Ban on *Ma'aseh Bereshit*: Sources, Contexts." In *Jewish and Christian Cosmogony in Late Antiquity*, edited by Lance Jenott and Sarit Kattan Gribetz, 39–63. Tübingen, Germany: Mohr Siebeck, 2013.

Gaard, Greta. "Toward a Queer Ecofeminism." *Hypatia* 12, no. 1 (1997): 114–37.
Garland-Thomson, Rosemarie. "The Case for Conserving Disability." *Bioethical Inquiry* 9 (2012): 339–55.
Garrard, Greg. "How Queer Is Green?" *Configurations* 18, no. 1–2 (2010): 73–96.
———. "Radical Pastoral?" *Studies in Romanticism* 35, no. 3 (1996): 449–65.
Goldschmidt, Rudolf Carel. *Paulinus' Churches at Nola: Texts, Translations, and Commentary*. Amsterdam: N.V. Noord-Hollandische Uitgevers Maatschappij, 1940.
Gregory of Nyssa. *Ascetical Works*. Translated by Virginia Woods Callahan. Fathers of the Church. Washington, DC: Catholic University of America Press, 1967.
———. *Vie de Sainte Macrine*. Edited by Pierre Maraval. Sources Chrétiennes 178. Paris: Cerf, 1971.
Gross-Diaz, Theresa. "Cat. 22: Tetradrachm (Coin) Portraying Queen Cleopatra VII, 37/33 B.C." In *Roman Art at the Art Institute of Chicago*, edited by Katharine A. Raff. Chicago: Art Institute of Chicago, 2017. https://publications.artic.edu/roman/reader/romanart/section/510/510_anchor, accessed June 4, 2017.
Grusin, Richard. "Introduction." In *The Nonhuman Turn*, edited by Richard Grusin, vii–xxix. Minneapolis: University of Minnesota Press, 2015.
Hadot, Pierre. *Philosophy as a Way of Life: Spiritual Exercises from Socrates to Foucault*. Edited by Arnold Davidson. Translated by Michael Chase. Oxford: Blackwell, 1995; French, 1981.
Hahn, Cynthia. *Strange Beauty: Issues in the Making and Meaning of Reliquaries, 400–circa 1204*. University Park: Pennsylvania State University Press, 2012.
———. "What Do Reliquaries Do for Relics?" *Numen* 57 (2010): 284–316.
Halberstam, Judith. *In a Queer Time and Place: Transgender Bodies, Subcultural Lives*. New York: New York University Press, 2005.
Haraway, Donna J. *The Companion Species Manifesto: Dogs, People, and Significant Others*. Chicago: University of Chicago Press, 2003.
———. *Simians, Cyborgs, and Women: The Reinvention of Nature*. New York: Routledge, 1991.
Harman, Graham. *Heidegger Explained: From Phenomenon to Thing*. Chicago: Open Court, 2007.
———. *Prince of Networks: Bruno Latour and Metaphysics*. Melbourne: Re.press, 2009.
———. *Tool-Being: Heidegger and the Metaphysics of Objects*. Chicago: Open Court, 2002.
———. "The Well-Wrought Broken Hammer: Object-Oriented Literary Criticism." *New Literary History* 43, no. 2 (2012): 183–203.
Harris, W. V. "Defining and Detecting Mediterranean Deforestation, 800 B.C.E to 700 C.E." In *The Ancient Mediterranean Environment Between Science and History*, edited by W. V. Harris. Leiden, Netherlands: Brill, 2013.
Harvey, Susan Ashbrook. "The Sense of a Stylite: Perspectives on Simeon the Elder." *Vigiliae Christianae* 42, no. 4 (1988): 376–94.
Hass, Petra. *Der* locus amoenus *in der antiken Literatur: Zu Theorie und Geschichte eines literarischen Motivs*. Bamberg, Germany: Wissenschaftler Verlag, 1998.
Hatuqa, Dalia. "Palestinian Farmers Thirsty for Exports." *Al Jazeera*, April 10, 2013. http://www.aljazeera.com/indepth/features/2013/04/201343126485592299.html.
Heidegger, Martin. "The Thing." In *Poetry, Language, Thought*, translated by Albert Hofstadter, 163–80. New York: Harper and Row, 1971.

Heidl, György. *Origen's Influence on the Young Augustine: A Chapter in the History of Origenism*. Piscataway, NJ: Gorgias, 2003.

Hess, Scott. "Postmodern Pastoral, Advertising, and the Masque of Technology." *Interdisciplinary Studies in Literature and Environment* 11 (2004): 71–100.

Hill, Leslie. *Maurice Blanchot and Fragmentary Writing: A Change of Epoch*. London: Continuum, 2012.

Hill, Leslie, and Helen Paris. *The Guerilla Guide to Performance Art*. London: Continuum, 2001.

Hodder, Ian. *Entangled: An Archaeology of the Relationships Between Humans and Things*. Chichester, UK: Wiley-Blackwell, 2012.

Hughes, J. Donald. *Pan's Travail: Environmental Problems of the Ancient Greeks and Romans*. Baltimore: Johns Hopkins University Press, 1994.

Hume, Angela. "Imagining Ecopoetics: An Interview with Robert Hass, Brenda Hillman, Evelyn Reilly, and Jonathan Skinner." *Interdisciplinary Studies in Literature and Environment* 19, no. 4 (2012): 751–66.

Ingold, Tim. *Making: Anthropology, Archaeology, Art and Architecture*. London: Routledge, 2013.

———. *The Perception of the Environment: Essays on Livelihood, Dwelling and Skill*. London: Routledge, 2000.

Irigaray, Luce. *An Ethics of Sexual Difference*. Translated by Carolyn Burke and Gillian C. Gill. Ithaca, NY: Cornell University Press, 1993.

———. *Speculum of the Other Woman*. Translated by Gillian C. Gill. Ithaca, NY: Cornell University Press, 1985; French ed., 1974.

Jenkins, Willis. "After Lynn White: Religious Ethics and Environmental Problems." *Journal of Religious Ethics* 37 (2009): 283–309.

Jensen, Robin M. *Baptismal Imagery in Early Christianity: Ritual, Visual, and Theological Dimensions*. Grand Rapids, MI: Baker Academic, 2012.

Jerome. "Life of Paul, the First Hermit." Translated by Paul B. Harvey. In *Ascetic Behavior in Greco-Roman Antiquity: A Sourcebook*, edited by Vincent L. Wimbush, 357–69. Minneapolis: Fortress, 1990.

Johnson, Alex. "How to Queer Ecology: One Goose at a Time." *Orion Magazine* 30, no. 2 (2011): 46–49.

Johnson, Dominic, ed. *Pleading in the Blood: The Art and Performances of Ron Athey*. Bristol, UK: Intellect, 2013.

Jones, Amelia. "Everybody Dies, Even the Gorgeous: Resurrecting the Work of Hannah Wilke." *Markszine* (2003).

Kearney, Richard. *The God Who May Be: A Hermeneutics of Religion*. Bloomington: Indiana University Press, 2001.

Keller, Catherine. *Cloud of the Impossible: Negative Theology and Planetary Entanglement*. New York: Columbia University Press, 2015.

———. *The Face of the Deep: A Theology of Becoming*. London: Routledge, 2003.

Kenney, John Peter. *The Mysticism of Saint Augustine: Rereading the* Confessions. New York: Routledge, 2005.

Kister, Menahem. "Tohu Wa-Bohu: Primordial Elements and *Creatio ex Nihilo*." *Jewish Studies Quarterly* 14 (2007): 229–56.

Kitchen Sisters and Lisa Morehouse. "Forbidding Fruit: How America Got Turned on to the Date." National Public Radio, *Morning Edition*, June 10, 2014. http://www.npr.org/

sections/thesalt/2014/06/10/320346869/forbidding-fruit-how-america-got-turned-on-to-the-date.

Kliszcz, Aneta, and Joanna Komorowska. "Glades of Dread: The Ecology and Aesthetics of *Loca Horrida*." In *Ecocriticism, Ecology, and the Cultures of Antiquity*, edited by Christopher Schliephake, 45–60. Lanham, MD: Lexington, 2017.

Knickerbocker, Scott. *Ecopoetics: The Language of Nature, the Nature of Language*. Amherst: University of Massachusetts Press, 2012.

Köckert, Charlotte. *Christliche Kosmologie und kaiserzeitliche Philosophie: Die Auslegung des Schöpfungsberichtes bei Origenes, Basilius und Gregor von Nyssa vor dem Hintergrund kaiserzeitlicher Timaeus-Interpretationen*. Tübingen, Germany: Mohr Siebeck, 2009.

Kohn, Eduardo. *How Forests Think: Toward an Anthropology Beyond the Human*. Berkeley: University of California Press, 2013.

Koltun-Fromm, Naomi. "Rock over Water: Pre-Historic Rocks and Primordial Waters from Creation to Salvation in Jerusalem." In *Jewish and Christian Cosmogony in Late Antiquity*, edited by Lance Jenott and Sarit Kattan Gribetz, 239–54. Tübingen, Germany: Mohr Siebeck, 2013.

Kouli, Maria. "Life of St. Mary of Egypt." In *Holy Women of Byzantium: Ten Saints' Lives in English Translation*, edited by Alice-Mary Talbot, 65–93. Washington, DC: Dumbarton Oaks Research Library and Collection, 1996.

Kötzsche-Breitenbruch, Lieselotte. "Zum Ring von Gregor von Nyssa." In *Tesserae: Festschrift für Josef Engemann*, edited by Ernst Dassmann, 291–98. Münster, Germany: Aschendorff, 1991.

Kristeva, Julia. *Powers of Horror: An Essay on Abjection*. Translated by Leon S. Roudiez. New York: Columbia University Press, 1982.

———. "Revolution in Poetic Language." In *The Kristeva Reader*, edited by Toril Moi, 89–136. Oxford: Basil Blackwell, 1986.

Krueger, Derek. "Liturgical Time and Holy Land Reliquaries in Early Byzantium." In *Saints and Sacred Matter: The Cult of Relics in Byzantium and Beyond*, edited by Cynthia Hahn and Holger A. Klein, 111–31. Washington, DC: Dumbarton Oaks, 2015.

———. "The Religion of Relics in Late Antiquity and Byzantium." In *Treasures of Heaven: Saints, Relics, and Devotion in Medieval Europe*, edited by Martina Bagnoli, Holger A. Klein, C. Griffith Mann, and James Robinson, 5–17. New Haven, CT: Yale University Press, 2010.

———. "Typological Figuration in Theodoret of Cyrrhus's Religious History and the Art of Postbiblical Narrative." *Journal of Early Christian Studies* 5, no. 3 (1997): 393–419.

———. *Writing and Holiness: The Practice of Authorship in the Early Christian East*. Divinations: Rereading Late Ancient Religion. Philadelphia: University of Pennsylvania Press, 2004.

Lane, Melissa. *Eco-Republic: What the Ancients Can Teach Us About Ethics, Virtue, and Sustainable Living*. Princeton, NJ: Princeton University Press, 2012.

Lehmann, Tomas. *Paulinus Nolanus und die Basilica Nova in Cimitile/Nola*. Wiesbaden, Germany: Reichert Verlag, 2004.

Leibovitch, Joseph. "The Reliquary Column of Dor." *Christian News from Israel* 5, no. 1–2 (1953): 22–23.

Leicht, Reimund. "Major Trends in Rabbinic Cosmology." In *Hekhalot Literature in Context: Between Byzantium and Babylonia*, edited by Ra'anon Boustan, Martha Himmelfarb, and Peter Schäfer, 245–78. Tübingen, Germany: Mohr Siebeck.

LeVasseur, Todd, and Anna Peterson. "Introduction." In *Religion and Ecological Crisis: The 'Lynn White Thesis' at Fifty*, edited by Todd LeVasseur and Anna Peterson, 1–17. New York: Routledge, 2017.

Lietzmann, H. *Das Leben des heiligen Symeon Stylites*. Texte und Untersuchung zur Geschichte der altchristlichen Literatur 32.4. Leipzig: Hinrichs, 1908.

Ling, Roger. *Ancient Mosaics*. Princeton, NJ: Princeton University Press, 1998.

Lingis, Alphonso. "Animal Body, Inhuman Face." In *Zoontologies: The Question of the Animal*, edited by Cary Wolfe, 165–82. Minneapolis: University of Minnesota Press, 2003.

Lyman, J. Rebecca. *Christology and Cosmology: Models of Divine Activity in Origen, Eusebius, and Athanasius*. Oxford Theological Monographs. Oxford: Clarendon, 1993.

MacKendrick, Karmen. *The Matter of Voice: Sensual Soundings*. New York: Fordham University Press, 2016.

Malm, Andreas, and Alf Hornborg. "The Geology of Mankind? A Critique of the Anthropocene Narrative." *Anthropocene Review* 1 (2014): 62–69.

Maritano, Mario. "Argomenti 'filosofici' di Origene contro la metensomatosi." In *Origeniana octava: Origen and the Alexandrian Tradition*, edited by Lorenzo Perrone, 503–35. Leuven: Peeters, 2002–3.

———. "L'argomentazione scritturistica di Origene contro i sostenitori della metensomatosi." In *Origeniana sexta: Origen and the Bible*, edited by Gilles Dorival and Alain Le Boulluec, 251–76. Leuven, Belgium: Peeters, 1995.

McKibben, Bill. "Foreword." In *Toward an Ecology of Transfiguration: Orthodox Christian Perspectives on Environment, Nature, and Creation*, edited by John Chryssavgis and Bruce V. Foltz, xiii–xvi. New York: Fordham University Press, 2013.

McKinnon, James W. "Liturgical Psalmody in the Sermons of St. Augustine: An Introduction." In *The Study of Medieval Chant: Paths and Bridges, East and West, in Honor of Kenneth Levy*, edited by Peter Jeffery, 7–24. Cambridge: Boydell, 2001.

Meijering, E. P. *Orthodoxy and Platonism in Athanasius: Synthesis or Antithesis?* Leiden, Netherlands: Brill, 1968.

Meloy, Ellen. *The Anthropology of Turquoise: Reflections on Desert, Sea, Stone, and Sky*. New York: Random House, 2002.

Miles, Margaret R. "'*Facie ad faciem*': Visuality, Desire, and the Discourse of the Other." *Journal of Religion* 87, no. 1 (2007): 43–58.

———. *Plotinus on Body and Beauty*. Oxford: Blackwell, 1999.

Miller, David L. "Theopoetry or Theopoetics?" *Crosscurrents* 60, no. 1 (March 2010): 6–23.

Miller, Patricia Cox. "Adam, Eve, and the Elephants: Asceticism and Animality." In *Ascetic Culture: Essays in Honor of Philip Rousseau*, edited by Blake Leyerle and Robin Darling Young, 253–68. Notre Dame, IN: University of Notre Dame Press, 2013.

———. *Biography in Late Antiquity: A Quest for the Holy Man*. Transformation of the Classical Heritage. Berkeley: University of California Press, 1983.

———. *The Corporeal Imagination: Signifying the Holy in Late Ancient Christianity*. Divinations: Rereading Late Ancient Religion. Philadelphia: University of Pennsylvania Press, 2009.

———. "Figuring Relics: A Poetics of Enshrinement." In *Saints and Sacred Matter: The Cult of Relics in Byzantium and Beyond*, edited by Cynthia Hahn and Holger A. Klein, 99–109. Washington, DC: Dumbarton Oaks, 2015.

———. "Is There a Harlot in This Text? Hagiography and the Grotesque." In *The Cultural Turn in Late Ancient Studies: Gender, Asceticism, and Historiography*, edited by

Dale B. Martin and Patricia Cox Miller, 88–102. Durham, NC: Duke University Press, 2005.

———. "Jerome's Centaur: A Hyper-Icon of the Desert." *Journal of Early Christian Studies* 4, no. 2 (1996): 209–33.

———. *The Poetry of Thought in Late Antiquity: Essays in Imagination and Religion*. Aldershot, UK: Ashgate, 2001.

Mortimer-Sandilands, Catriona. "Unnatural Passions? Notes Toward a Queer Ecology." *Invisible Culture: An Electronic Journal for Visual Culture*, no. 9 (2005). https://www.rochester.edu/in_visible_culture/Issue_9/sandilands.html.

Morton, Timothy. *Dark Ecology: For a Logic of Future Coexistence*. New York: Columbia University Press, 2016.

———. *The Ecological Thought*. Cambridge, MA: Harvard University Press, 2010.

———. *Ecology Without Nature: Rethinking Environmental Aesthetics*. Cambridge, MA: Harvard University Press, 2009.

———. "From Modernity to the Anthropocene: Ecology and Art in the Age of Asymmetry." *International Social Science Journal* 63, no. 207/208 (2012): 39–51.

———. "Here Comes Everything: The Promise of Objected-Oriented Ontology." *Qui Parle: Critical Humanities and Social Sciences* 19 (2011): 163–90.

———. *Hyperobjects: Philosophy and Ecology After the End of the World*. Minneapolis: University of Minnesota Press, 2013.

———. "Queer Ecology." *Proceedings of the Modern Language Association* 125, no. 2 (2010): 273–82.

Nancy, Jean-Luc. *The Inoperative Community*. Edited by Peter Connor. Translated by Peter Connor, Lisa Garbus, Michael Holland, and Simona Sawhney. Theory and History of Literature, vol. 76. Minneapolis: University of Minnesota Press, 1991.

———. *Listening*. Translated by Charlotte Mandell. New York: Fordham University Press, 2007.

Neusner, Jacob. Genesis Rabbah, *the Judaic Commentary to the Book of Genesis: A New American Translation, Volume I*. Atlanta: Scholars Press, 1985.

Newman, Lance, and Laura Walls. "Cosmopolitics and the Radical Pastoral: A Conversation with Lawrence Buell, Hsuan Hsu, Antony Lioi, and Paul Outka." *Journal of Ecocriticism* 3, no. 2 (2011): 58–71.

Niehoff, Maren. "*Creatio ex Nihilo* Theology in *Genesis Rabbah* in Light of Christian Exegesis." *Harvard Theological Review* 99, no. 1 (2005): 37–64.

———. "Did the *Timaeus* Create a Textual Community?" *Greek, Roman, and Byzantine Studies* 47 (2007): 161–91.

Nightingale, Andrea. *Once Out of Nature: Augustine on Time and the Body*. Chicago: University of Chicago Press, 2011.

Northcott, Michael S. "Lynn White Jr. Right and Wrong: The Anti-Ecological Character of Latin Christianity and the Pro-Ecological Turn of Protestantism." In *Religion and Ecological Crisis: The "Lynn White Thesis" at Fifty*, edited by Todd LeVasseur and Anna Peterson, 61–74. New York: Routledge, 2017.

O'Brien, Carl Séan. *The Demiurge in Ancient Thought: Secondary Gods and Divine Mediators*. Cambridge: Cambridge University Press, 2015.

Oldfather, W., et al. *Studies in the Text Traditions of St. Jerome's* Vitae patrum. Urbana: University of Illinois Press, 1943.

Oord, Thomas Jay, ed. *Theologies of Creation:* Creatio ex Nihilo *and Its New Rivals*. New York: Routledge, 2015.

Origen. *Contra Celsum*. Translated by Henry Chadwick. Cambridge: Cambridge University Press, 1965.

———. *Contre Celse*. Edited by M. Borret. Sources Chrétiennes 132, 136, 147, 150, 227. Paris: Cerf, 1967–76.

———. *Die Kommentierung des Buches Genesis*. Edited and translated by Karin Metzler. Berlin: De Gruyter, 2010.

———. *On First Principles*. Translated by G. W. Butterworth. New York: Harper and Row, 1966.

———. *Traité des principes*. Edited by Henry Crouzel and Manlio Simonetti. Sources Chrétiennes 252, 253, 268, 269, 312. Paris: Cerf, 1978–84.

Papalexandrou, Amy. "Text in Context: Eloquent Monuments and the Byzantine Beholder." *Word and Image* 17, no. 3 (2001): 259–83.

Parani, Maria. "Byzantine Bridal Costume." In *Δώρημα: A Tribute to the A. G. Leventis Foundation on the Occasion of Its 20th Anniversary*, 185–216. Nicosia, Cyprus: A. G. Leventis Foundation, 2000.

Parker, Holt. "Women and Medicine." In *A Companion to Women in the Ancient World*, edited by Sharon L. James and Sheila Dillon, 107–24. Chichester, UK: Wiley Blackwell, 2012.

Peers, Glenn. "Adam's Anthropocene." *Postmedieval: A Journal of Medieval Cultural Studies* (2016), 1–11.

———. "Byzantine Things in the World." In *Byzantine Things in the World*, edited by Glenn Peers, 41–84. New Haven, CT: Yale University Press, 2013.

———. "Crosses' Work Underfoot: Christian Spolia in the Late Ancient Mosque at Shivta in the Negev Desert (Israel)." *Eastern Christian Art* 8 (2011): 101–19.

———. "Late Antique Making and Wonder," forthcoming.

———. "Object Relations: Theorizing the Late Antique Viewer." In *The Oxford Handbook of Late Antiquity*, edited by Scott Fitzgerald Johnson, 970–87. Oxford: Oxford University Press, 2012.

———. *Sacred Shock: Framing Visual Experience in Byzantium*. University Park: Pennsylvania State University Press, 2004.

———. "Sense Lives of Byzantine Things." In *Knowing Bodies, Passionate Souls: Sense Perceptions in Byzantium*, edited by Susan Ashbrook Harvey and Margaret Mullett, 11–30. Washington, DC: Dumbarton Oaks, 2017.

Perl, Eric Justin David. "Symbol, Sacrament, and Hierarchy in Saint Dionysios the Areopagite." *Greek Orthodoxy Theological Review* 39, no. 3/4 (1994): 311–56.

Perrone, Lorenzo. *Origenes XIII: Die neuen Psalmenhomilien: Eine kritische Edition des Codex Monacensis Graecus 314*. Berlin: Walter de Gruyter, 2015.

Phillips, Adam. *Darwin's Worms: On Life Stories and Death Stories*. New York: Basic, 2000.

Philo. *On the Creation: Allegorical Interpretation of Genesis 2 and 3*. Translated by F. H. Colson and G. H. Whitaker. Loeb Classical Library 226. Cambridge, MA: Harvard University Press, 1929.

Plato. *Timaeus, Critias, Cleitophon, Menexenus, Epistles*. Translated by R. G. Bury. Loeb Classical Library 234. Cambridge, MA: Harvard University Press, 1929.

Plotinus. *Enneads, Volume I: Porphyry on the Life of Plotinus—Ennead I*. Translated by A. H. Armstrong. Loeb Classical Library 440. Cambridge, MA: Harvard University Press, 1969.

Procopius of Gaza et al. *Opera omnia, tomus tertius*. Patrologia Graeca 87.3, 1865.

Rasch, Jürgen J., and Achim Arbeiter. *Das Mausoleum der Constantina in Rom*. Mainz, Germany: Verlag Philipp von Zabern, 2007.

Reed, Annette Yoshiko. "From 'Pre-Emptive Exegesis' to 'Pre-Emptive Speculation'? *Ma'aseh Bereshit* in *Genesis Rabbah* and *Pirqei DeRabbi Eliezer*." In *With Letters of Light/Otiyot Shel or: Studies in Early Jewish Apocalypticism and Mysticism in Honour of Rachel Elior*, edited by D. Arbel and A. Orlov, 115–32. Berlin: De Gruyter, 2010.

———. "Reading Augustine and/as Midrash: Genesis 6 in *Genesis Rabbah* and the *City of God*." In *Midrash and Context*, edited by Lieve M. Teugels and Rivka Ulmer, 61–110. Piscataway, NJ: Gorgias, 2007.

Riley, Matthew T. "A Spiritual Democracy of All God's Creatures: Ecotheology and the Animals of Lynn White Jr." In *Divinanimality: Animal Theory, Creaturely Theology*, edited by Stephen D. Moore, 241–60. New York: Fordham University Press, 2014.

Rist, John M. "A Note on Eros and Agape in Pseudo-Dionysius." *Vigiliae Christianae* 20 (1966): 235–43.

Rivera, Mayra. *Poetics of the Flesh*. Durham, NC: Duke University Press, 2015.

Roach, John. "'Methuselah' Palm Grown from 2,000-Year-Old Seed Is a Father." *National Geographic*, March 24, 2015. http://news.nationalgeographic.com/2015/03/150324-ancient-methuselah-date-palm-sprout-science/.

Roughgarden, Joan. *Evolution's Rainbow: Diversity, Gender, and Sexuality in Nature and People*. Berkeley: University of California Press, 2004.

Rousseau, Philip. "Human Nature and Its Material Setting in Basil of Caesarea's Sermons on the Creation." *Heythrop Journal* 49 (2008): 222–39.

Ruaro, Enrica. "God and the Worm: The Twofold Otherness in Pseudo-Dionysius's Theory of Dissimilar Images." *American Catholic Philosophical Quarterly* 82, no. 4 (2008): 581–92.

———. "'Resurrection: The Hope of Worms': The Dispute Between Celsus and Origen on the Resurrection of the Body." In *Perspectives sur le néoplatonisme: International Society of Neoplatonic Studies: Actes du colloque de 2006*, edited by Martin Achard, Wayne Hankey, and Jean-Marc Narbonne, 111–22. Québec: Presses de l'Université Laval, 2009.

Rubenstein, Mary-Jane. "Cosmic Singularities: On the Nothing and the Sovereign." *Journal of the American Academy of Religion* 80, no. 2 (2012): 485–517.

———. "Dionysius, Derrida, and the Critique of 'Ontotheology.'" *Modern Theology* 24, no. 4 (2008): 725–41.

———. "The Lure of Pan(en)theism: Difference and Desire in *Divine Enticement*." *Theology and Sexuality* 18, no. 2 (2012): 113–17.

Runia, David T. "Philo and the Early Christian Fathers." In *The Cambridge Companion to Philo*, edited by Adam Kamesar, 210–30. Cambridge: Cambridge University Press, 2009.

———. *Philo of Alexandria and the* Timaeus *of Plato*. Leiden, Netherlands: Brill, 1986.

———. *Philo of Alexandria, On the Creation of the Cosmos According to Moses: Introduction, Translation, and Commentary*. Leiden, Netherlands: Brill, 2001.

Sallis, John. *Chorology: On Beginning in Plato's* Timaeus. Bloomington: Indiana University Press, 1999.

———. *Platonic Legacies*. SUNY Series in Contemporary Continental Philosophy. Albany: State University of New York Press, 2004.

Sandilands, Catriona. "Lavender's Green? Some Thoughts on Queer(y)ing Environmental Politics." *Undercurrents* 6 (1994): 20–25.

———. "Queer Life? Ecocriticism After the Fire." In *The Oxford Handbook of Ecocriticism*, edited by Greg Garrard, 305–19. Oxford: Oxford University Press, 2014.

Santmire, H. Paul. *The Travail of Nature: The Ambiguous Ecological Promise of Christian Theology*. Philadelphia: Fortress, 1985.

Santner, Eric L. *On Creaturely Life: Rilke, Benjamin, Sebald*. Chicago: University of Chicago Press, 2006.

Schaefer, Donovan O. *Religious Affects: Animality, Evolution, and Power*. Durham, NC: Duke University Press, 2015.

Schäfer, Peter. "Bereshit Bara Elohim: Bereshit Rabba, Parashah 1, Reconsidered." In *Empsychoi Logoi—Religious Innovations in Antiquity: Studies in Honour of Pieter Willem van der Horst*, edited by Alberdina Houtman, Albert de Jong, and Magdalena Wilhelmina Misset-van de Weg, 267–89. Leiden, Netherlands: Brill, 2008.

Schliephake, Christopher. "Introduction." In *Ecocriticism, Ecology, and the Cultures of Antiquity*, edited by Christopher Schliephake, 1–15. Lanham, MD: Lexington, 2017.

Scholem, Gershom. *Jewish Gnosticism, Merkavah Mysticism, and Talmudic Tradition*. 2nd rev. ed. New York: Jewish Theological Seminary, 1965.

Şen, Zekai. *Wadi Hydrology*. Boca Raton, FL: CRC Press, 2008.

Seymour, Nicole. *Strange Natures: Futurity, Empathy, and the Queer Ecological Imagination*. Champaign: University of Illinois Press, 2013.

Shaviro, Steven. *The Universe of Things: On Speculative Realism*. Minneapolis: University of Minnesota Press, 2014.

Sheldon, Rebekah. "Form/Matter/Chora: Object-Oriented Ontology and Feminist New Materialism." In *The Nonhuman Turn*, edited by Richard Grusin, 193–222. Minneapolis: University of Minnesota Press, 2015.

Shuttleton, David. "The Queer Politics of Gay Pastoral." In *De-Centering Sexualities: Politics and Representations Beyond the Metropolis*, edited by Richard Phillips, Diane Watt, and David Shuttleton, 123–42. London: Routledge, 2000.

Siebers, Tobin. *Disability Aesthetics*. Ann Arbor: University of Michigan Press, 2010.

Siewers, Alfred K. *Strange Beauty: Ecocritical Approaches to Early Medieval Landscape*. New York: Palgrave Macmillan, 2009.

Skinner, Jonathan. "Ecopoetics." *Jacket2* (2011).

Smith, J. Warren. "Divine Ecstasy and Divine Simplicity: The Eros Motif in Pseudo-Dionysius's Soteriology." *Pro Ecclesia* 21, no. 2 (2012): 211–27.

Soskice, Janet Martin, ed. "Creation 'ex Nihilo' and Modern Theology." *Modern Theology* 29, no. 2 (2013).

Spier, Jeffrey. "Some Unconventional Early Byzantine Rings." In *"Intelligible Beauty": Recent Research on Byzantine Jewelry*, edited by Chris Entwistle and Noël Adams, 13–19. London: British Museum, 2010.

Spinks, Bryan D. *The Sanctus in the Eucharistic Prayer*. Cambridge: Cambridge University Press, 1991.

Spittler, Janet E. *Animals in the Apocryphal Acts of Apostles*. Tübingen, Germany: Mohr Siebeck, 2008.

Stang, Charles M. *Apophasis and Pseudonymity in Dionysius the Areopagite: "No Longer I"*. Oxford: Oxford University Press, 2012.

———. "The Beginning and End of All Hierarchy." In *The Open Body: Essays in Anglican Ecclesiology*, edited by Zachary Guiliano and Charles M. Stang, 103–24. New York: Peter Lang, 2012.

Stanley, David J. "New Discoveries at Santa Costanza." *Dumbarton Oaks Papers* 28 (1994): 257–61.

Strack, H. L., and Günter Stemberger. *Introduction to the Talmud and Midrash*. 2nd ed. Translated and edited by Markus Bockmuehl. Minneapolis: Fortress, 1996.

Talgam, Rina. *Mosaics of Faith: Floors of Pagans, Jews, Samaritans, Christians, and Muslims in the Holy Land*. Jerusalem and University Park: Yad Ben-Zvi Press and Pennsylvania State University Press, 2014.

Tanner, Kathryn. "*Creatio ex Nihilo* as Mixed Metaphor." *Modern Theology* 29, no. 2 (2013): 138–55.

Tengberg, M. "Beginnings and Early History of Date Palm Garden Cultivation in the Middle East." *Journal of Arid Environments* 86 (2012): 139–47.

Terian, Abraham. *Philonis Alexandrini de animalibus: The Armenian Text with an Introduction, Translation, and Commentary*. Chico, CA: Scholars Press, 1981.

van den Hoek, Annawies. "Assessing Philo's Influence in Christian Alexandria: The Case of Origen." In *Shem in the Tents of Japhet: Essays on the Encounter of Judaism and Hellenism*, edited by James L. Kugel, 223–39. Leiden, Netherlands: Brill, 2002.

———. "Philo and Origen: A Descriptive Catalogue of Their Relationship." *Studia Philonica Annual* (2000), 44–121.

van den Hoek, Annewies, and John Joseph Hermann Jr. *Pottery, Pavements, and Paradise: Iconographic and Textual Studies on Late Antiquity*. Leiden, Netherlands: Brill, 2013.

van Winden, J. C. M. "'An Appropriate Beginning': The Opening Passage of Saint Basil's *In Hexaemeron*." *Revue de l'histoire des religions* 203 (1983): 307–11.

———. *Calcidius on Matter: His Doctrine and Sources—A Chapter in the History of Platonism*. Leiden, Netherlands: Brill, 1965.

———. "'Terra autem stupida quadam erat admiratione': Reflections on a Remarkable Translation of Genesis 1:2a." In *Arche: A Collection of Patristic Studies by J. C. M. van Winden*, edited by J. den Boeft and D. T. Runia, 107–15. Leiden, Netherlands: Brill, 1977.

Vasey, Paul, and Volker Sommer, eds. *Homosexual Behaviour in Animals: An Evolutionary Perspective*. Cambridge: Cambridge University Press, 2006.

Vikan, Gary. "Art and Marriage in Early Byzantium." *Dumbarton Oaks Papers* 44 (1990): 145–63.

Vorwerk, Matthias. "Maker or Father? The Demiurge from Plutarch to Plotinus." In *One Book, the Whole Universe: Plato's Timaeus Today*, edited by Richard D. Mohr and Barbara M. Sattler, 79–100. Los Vegas: Parmenides, 2010.

Waliszewski, Tomasz. "Mosaics." In *The Petra Church*, 219–70. Amman, Jordan: American Center of Oriental Research, 2001.

Warr, Tracey, ed. *The Artist's Body*. London: Phaidon, 2000.

Weingarten, Susan. *The Saint's Saints: Hagiography and Geography in Jerome*. Leiden, Netherlands: Brill, 2005.

Wharton, Annabel Jane. *Architectural Agents: The Delusional, Abusive, Addictive Lives of Buildings*. Minneapolis: University of Minnesota Press, 2015.

Whitaker, Iain S., et al. "Larval Therapy from Antiquity to the Present Day: Mechanisms of Action, Clinical Applications and Future Potential." *Postgraduate Medical Journal* 83 (2007): 409–13.

White, Lynn Jr. "The Historical Roots of Our Ecological Crisis." *Science* 155, no. 3767 (1967): 1203–7.

Whitehead, Alfred North. *Process and Reality*. New York: Free Press, 1978; orig. 1929.

Wilson, Edward O. "Biophilia and the Conservation Ethic." In *The Biophilia Hypothesis*, edited by Stephen R. Kellert and Edward O. Wilson, 31–41. Washington, DC: Island, 1993.

Winnubst, Shannon. *Queering Freedom*. Bloomington: Indiana University Press, 2006.

Wolfe, Cary. "Introduction: What Is Posthumanism?" In *What Is Posthumanism?* edited by Cary Wolfe, xi–xxxiv. Minneapolis: University of Minnesota Press, 2010.

Wyschogrod, Edith. *Saints and Postmodernism: Revisioning Moral Philosophy*. Chicago: University of Chicago Press, 1990.

Yasin, Ann Marie. "Prayers on Site: The Materiality of Devotional Graffiti and the Production of Early Christian Sacred Space." In *Viewing Inscriptions in the Late Antique and Medieval World*, edited by Antony Eastmond, 36–60. New York: Cambridge University Press, 2015.

Zourabichvili, François. "Six Notes on the Percept (on the Relation Between the Critical and the Clinical)." In *Deleuze: A Critical Reader*, edited by Paul Patton, 188–216. Cambridge, MA: Blackwell, 1996.

Index